Biomedical Applications of Computer Modeling

Pharmacology and Toxicology: Basic and Clinical Aspects

Mannfred A. Hollinger, Series Editor
University of California, Davis

Published Titles

Manual of Immunological Methods, 1999, Pauline Brousseau, Yves Payette, Helen Tryphonas, Barry Blakley, Herman Boermans, Denis Flipo, Michel Fournier

CNS Injuries: Cellular Responses and Pharmacological Strategies, 1999, Martin Berry and Ann Logan

*Infectious Diseases in Immunocompromised Hosts,*1998, Vassil St. Georgiev

Pharmacology of Antimuscarinic Agents, 1998, Laszlo Gyermek

Basis of Toxicity Testing, Second Edition, 1997, Donald J. Ecobichon

Anabolic Treatments for Osteoporosis, 1997, James F. Whitfield and Paul Morley

Antibody Therapeutics, 1997, William J. Harris and John R. Adair

Muscarinic Receptor Subtypes in Smooth Muscle, 1997, Richard M. Eglen

Antisense Oligodeonucleotides as Novel Pharmacological Therapeutic Agents, 1997, Benjamin Weiss

Airway Wall Remodelling in Asthma, 1996, A.G. Stewart

Drug Delivery Systems, 1996, Vasant V. Ranade and Mannfred A. Hollinger

Brain Mechanisms and Psychotropic Drugs, 1996, Andrius Baskys and Gary Remington

Receptor Dynamics in Neural Development, 1996, Christopher A. Shaw

Ryanodine Receptors, 1996, Vincenzo Sorrentino

Therapeutic Modulation of Cytokines, 1996, M.W. Bodmer and Brian Henderson

Pharmacology in Exercise and Sport, 1996, Satu M. Somani

Placental Pharmacology, 1996, B. V. Rama Sastry

Pharmacological Effects of Ethanol on the Nervous System, 1996, Richard A. Deitrich

Immunopharmaceuticals, 1996, Edward S. Kimball

Chemoattractant Ligands and Their Receptors, 1996, Richard Horuk

Pharmacological Regulation of Gene Expression in the CNS, 1996, Kalpana Merchant

Experimental Models of Mucosal Inflammation, 1995, Timothy S. Gaginella

Human Growth Hormone Pharmacology: Basic and Clinical Aspects, 1995, Kathleen T. Shiverick and Arlan Rosenbloom

Placental Toxicology, 1995, B. V. Rama Sastry

Stealth Liposomes, 1995, Danilo Lasic and Frank Martin

TAXOL®: Science and Applications, 1995, Matthew Suffness

Endothelin Receptors: From the Gene to the Human, 1995, Robert R. Ruffolo, Jr.

Alternative Methodologies for the Safety Evaluation of Chemicals in the Cosmetic Industry,1995, Nicola Loprieno

Phospholipase A_2 in Clinical Inflammation: Molecular Approaches to Pathophysiology, 1995, Keith B. Glaser and Peter Vadas

Serotonin and Gastrointestinal Function, 1995, Timothy S. Gaginella and James J. Galligan

Chemical and Structural Approaches to Rational Drug Design, 1994, David B. Weiner and William V. Williams

Biological Approaches to Rational Drug Design, 1994, David B. Weiner and William V. Williams

Direct and Allosteric Control of Glutamate Receptors, 1994, M. Palfreyman, I. Reynolds, and P. Skolnick

Genomic and Non-Genomic Effects of Aldosterone, 1994, Martin Wehling

Pharmacology and Toxicology: Basic and Clinical Aspects

Published Titles (*Continued*)

Peroxisome Proliferators: Unique Inducers of Drug-Metabolizing Enzymes, 1994, David E. Moody

Angiotensin II Receptors, Volume I: Molecular Biology, Biochemistry, Pharmacology, and Clinical Perspectives, 1994, Robert R. Ruffolo, Jr.

Angiotensin II Receptors, Volume II: Medicinal Chemistry, 1994, Robert R. Ruffolo, Jr.

Beneficial and Toxic Effects of Aspirin, 1993, Susan E. Feinman

Preclinical and Clinical Modulation of Anticancer Drugs, 1993, Kenneth D. Tew, Peter Houghton, and Janet Houghton

In Vitro *Methods of Toxicology*, 1992, Ronald R. Watson

Human Drug Metabolism from Molecular Biology to Man, 1992, Elizabeth Jeffreys

Platelet Activating Factor Receptor: Signal Mechanisms and Molecular Biology, 1992, Shivendra D. Shukla

Biopharmaceutics of Ocular Drug Delivery, 1992, Peter Edman

Pharmacology of the Skin, 1991, Hasan Mukhtar

Inflammatory Cells and Mediators in Bronchial Asthma, 1990, Devendra K. Agrawal and Robert G. Townley

Biomedical Applications of Computer Modeling

Edited by

Arthur Christopoulos

CRC Press
Boca Raton London New York Washington, D.C.

Library of Congress Cataloging-in-Publication Data

Biomedical applications of computer modeling / [edited by] Arthur Christopoulos.
p. cm. — (Pharmacology and toxicology)
Includes bibliographical references and index.
ISBN 0-8493-0100-9
1. Medicine—Computer simulation. 2. Medical sciences—Computer simulation. 3. Biological models—Computer simulation. I. Christopoulos, Arthur. II. Pharmacology & toxicology (Boca Raton, Fla.)
[DNLM: 1. Computer Simulation. 2. Models, Biological. QH 324.8 B6152 2000]
R859.7.C65 B54 2000
610'.1'13—dc21 00-044445
CIP

Direct all inquiries to CRC Press LLC, 2000 N.W. Corporate Blvd., Boca Raton, Florida 33431.

International Standard Book Number 0-8493-0100-9
Library of Congress Card Number 00-044445
Printed in the United States of America 1 2 3 4 5 6 7 8 9 0
Printed on acid-free paper

Preface

Great advances are constantly being made in the study of the molecular components that comprise many biological responses. However, the purpose of biological research is not only to delineate these processes, but also to understand how they work together. To this end, the modeling of biological processes represents one of the most powerful quantitative tools applied to the study of complicated systems.

What is a model? In the simplest sense, a model is a "facsimile" of reality. Models can never prove that a biological hypothesis is correct, but they can be a guide to a deeper understanding of a researcher's experimental system and towards better experimental design. Until the last decade or so, most general modeling approaches have required significant amounts of computer time, expertise, and resources, as well as highly specialized and (often) custom-written programs. As a consequence, most of the literature that is currently available regarding computer modeling and its applications in biology has its roots in the early computer era, is enshrined in mathematical jargon, and generally remains underutilized by the majority of biomedical researchers.

Nowadays, the bar has been lowered. Desktop and notebook microcomputers are practically ubiquitous in most biomedical environments and many common software packages have evolved to such an extent that most researchers have at their disposal a means for easily addressing many of their data analysis problems. Furthermore, these same computer packages are readily applicable to the modeling process, quickly performing calculations that would have taken days on the computers of yesteryear. However, most of these software/hardware features are either undocumented or unappreciated, and it is therefore not surprising that an increasing number of scientific papers are now appearing in the literature that describe how various readily available and user-friendly computer packages may be exploited in the data analysis and modeling process. The stage is now set for a more widespread appreciation of what modeling can offer.

This book brings together contributions from experts in the field that encompass different types of computer modeling approaches in the hope of providing a convenient source of readily applicable modeling information. Included are examples of equilibrium modeling, dynamic/kinetic modeling, and stochastic modeling, as well as overviews of the fitting of models to experimental data and of computer-intensive statistical analyses.

In general, a successful model is one that provides insight into mechanisms, allows researchers to design better experiments in order to test a given hypothesis, and dissuades them from too much wishful thinking. This book is specifically targeted to those interested in the modeling of biological processes, and it is hoped that the examples contained within prove useful in illustrating the place of computer modeling in biomedical research.

Arthur Christopoulos

Dedication

To Tracie

Editor

Arthur Christopoulos graduated from the Victorian College of Pharmacy, Monash University, Victoria, Australia in 1990 and obtained his Ph.D. in Pharmacology in 1997. He subsequently gained a position as a postdoctoral associate in the Department of Psychiatry, University of Minnesota, in the United States, prior to taking up his present position as a research fellow in the Department of Pharmacology, University of Melbourne, Victoria, Australia in 1999. His research interests include the study of novel modes of G protein-coupled receptors regulation and the application of computer modeling to analytical pharmacology. He is a member of the American Society of Pharmacology and Experimental Therapeutics, the Society for Neuroscience, the International Brain Research Organization, the New York Academy of Sciences, the Australasian Society of Clinical and Experimental Pharmacology and Toxicology, and the Pharmaceutical Society of Australia.

Contributors

Tommaso Costa
Department of Pharmacology
Istituto Superiore di Sanità
Rome, Italy

Arthur Christopoulos
Department of Pharmacology
University of Melbourne
Parkville, Victoria, Australia

Frederick J. Ehlert
Department of Pharmacology
College of Medicine
University of California
Irvine, California

Terry Kenakin
Department of Receptor Biochemistry
Glaxo Wellcome Research and Development
Research Triangle Park, North Carolina

Michael J. Lew
Department of Pharmacology
University of Melbourne
Parkville, Victoria, Australia

Jennifer J. Linderman
Department of Chemical Engineering
University of Michigan
Ann Arbor, Michigan

John Ludbrook
Professor (Retired)
University of Melbourne
North Carlton, Victoria, Australia

Michael W. Lutz
Section Head, Cheminformatics
Glaxo Wellcome, Inc.
Research Triangle Park, North Carolina

H. Ongun Onaran
Department of Pharmacology and Clinical Pharmacology
Ankara University Faculty of Medicine
Ankara, Turkey

Peter J. Woolf
Department of Chemical Engineering
University of Michigan
Ann Arbor, Michigan

Table of Contents

Acknowledgments

I would like to express my gratitude to Prof. James A. Angus (University of Melbourne), Prof. Esam E. El-Fakahany (University of Minnesota), and Dr. Fred Mitchelson (Victorian College of Pharmacy, Monash University) for their contributions to and ongoing support of my interests in computer modeling and analytical pharmacology. I am particularly indebted to Dr. Michael J. Lew (University of Melbourne) for his assistance and constructive criticisms throughout the preparation of this book, and to all the authors for their timely and insightful contributions. Finally, I would like to thank Dr. Tracie L. Pierce for her continued support, Prof. Mannfred Hollinger (University of California, Davis) for his guidance as series editor, and the expert professional assistance of Becky McEldowney and all at CRC Press.

1 The Pharmacologic Consequences of Modeling Synoptic Receptor Systems

Terry Kenakin

CONTENTS

1.1 INTRODUCTION

This chapter will discuss the use of mathematical models in synoptic receptor systems. The term synoptic refers to the comprehensive nature of G protein-coupled receptor systems since, in many cases, the receptor, G protein, and stimulus-response mechanisms are integral parts of the model and must be considered in the equations.

Probably the most important reason to formulate a pharmacologic model of a receptor interaction is to construct a tool with which one can go on to predict the results of experiments not yet conducted. A second advantage of such models is to make obvious the next experiment to be done. A potential hazard to be avoided is the bias that can sometimes result from the act of creating a model which can, in turn, result in a mistaken belief that the model accurately reflects nature. At best, a stand-alone model of drug–receptor interaction is a module within a vastly complicated set of modules of interactive biological units. In practice, models should continually be tuned in accordance with the new experimental data used to evaluate them. This chapter will present selected examples of where modeling can make a difference in the thought process aimed at understanding the interaction between drugs and receptors.

Essentially all experimenters construct intuitive models on a daily basis; an experiment is designed, conducted, and the numeric data compiled. From this, an image of the biology is made. By adding quantitative behavior to the model, it

0-8493-0100-9/01/$0.00+$.50

TABLE 1.1
Types of Mathematical Models

Model Type	Complexity	Estimability of Parameters
Caricature	Low	Low
Simulation	High	High
Parsimonious	Low	High
Heuristic	High	Low

becomes immensely more powerful. Thus, rules of behavior are defined mathematically and quantitative experiments are designed to test those rules. The rules become further defined and the model is enriched until it may offer insights into the complex biology being observed.

There are many types of models and they can generally be classified in terms of the parameters used to describe them, specifically how many of them there are and how estimable they are. Thus, complex models have many parameters and simple ones, few. Models can be characterized in terms of parameter estimability. If the model is describing a relatively unknown system where correspondingly little is known about the value of the parameters (little can be estimated), then the model becomes a caricature of the system. These models are cartoons of the system, but they still may guide speculation. At the opposite end of this scale are models that encompass a large amount of knowledge about the system with a large number of parameters that can be estimated. These are simulation models that can be used to predict the "reality" that the system behavior demonstrates under different circumstances.

Models can also be described in terms of their complexity. Very simple models that utilize the barest minimum in numbers of parameters are referred to as parsimonious models. For these models, the estimability of the parameter becomes a test for whether it should be included in the model or not. On the opposite end of this scale are heuristic models. These contain a large number of parameters, many of which probably cannot be estimated independently. These models capture the behavior of complex systems and can be used to predict behavior not intuitively obvious. In general, these model types are summarized in Table 1.1. The following will discuss how models can be used to optimize systems, predict complex behavior of systems, and be used to design more explicit models; specific examples from receptor pharmacology will be used demonstrate this approach.

1.2 MODELS CAN BE USED TO TROUBLESHOOT AND OPTIMIZE EXPERIMENTAL SYSTEMS

Models can be used to determine limits, capabilities, and weaknesses of systems. Importantly, they can also define how to optimize complex systems. For example, it is known that an HIV viral coat protein, gp120, a single transmembrane cellular protein named CD4, and the seven transmembrane chemokine receptor CXCR4 form a ternary complex on the cell surface to allow membrane fusion and subsequent

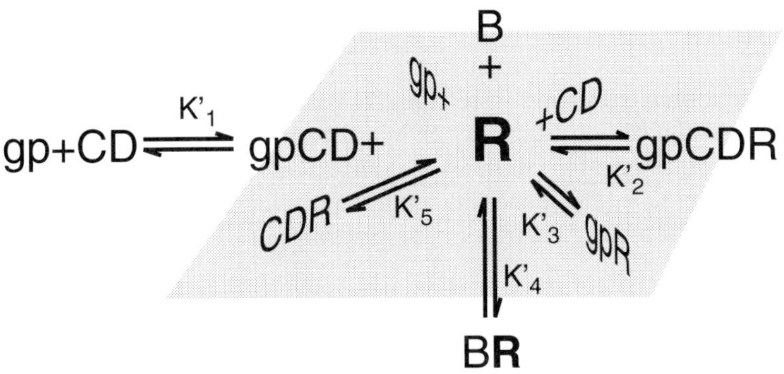

FIGURE 1.1 Model depicting the interaction of HIV viral coat protein gp120, and membrane bound proteins CD4, and the chemokine receptor CXCR4 forming a ternary complex (gpCDR) as a prelude to HIV infection. B refers to an antagonist that prevents formation of the ternary complex between the receptor, CD4, and gp120; K_1' through K_5' refer to equilibrium association constants of the receptor complexes.

HIV infection of healthy cells to occur.[1] A system that could quantify this interaction would be valuable in the study of drugs designed to block it. The model in Figure 1.1 can be used; it consists of the viral coat protein gp120 (the concentration of which is denoted [gp]), the membrane protein CD4 (denoted [CD]), the membrane bound receptor CXCR4 (denoted [R]), and a ligand that binds to CXCR4 (denoted [B]). For membrane fusion (and subsequent entry of the HIV genome into healthy cells), the three proteins gp120, CD4, and CXCR4 must form a ternary complex. The ligand B, by virtue of binding to CXCR4 and resulting preclusion of ternary complex formation, blocks the entry of HIV into uninfected cells.

A scintillation proximity assay (SPA) for screening such ligands could be made by membrane bound CXCR4 on SPA beads and the addition of soluble gp120 and radioactive CD4 (commercially available). A practical problem in this approach is the prohibitive cost of purified gp120. However, broths containing crude fractions of gp120 are readily available. Therefore, the practical problem to be overcome is to gauge the extent to which variation in the true concentration of gp120 in crude broths will affect the observed affinities of possible anti-HIV ligands while the SPA is being conducted.

This model quantifies the effects of unknown variations of gp120 in supernatants used in screening on the ability of the screen to detect ligands (Figure 1.1). It assumes no prohibited interactions (i.e., all proteins involved may interact with each other). The CXCR4 receptor conservation equation is given as:

$$[R_{total}] = [R] + [CDR] + [gpCDR] + [gpR] + [BR] \tag{1}$$

The SPA signal is generated by radioactive CD4 in proximity with the bead that could emanate from the two species, [gpCDR] and [CDR].

The signal, as a fraction of the total receptor concentration is given by:

$$\text{Fractional signal} = \wp = \frac{[\text{gpCDR}] + [\text{CDR}]}{[\text{R}_{\text{total}}]} \tag{2}$$

From the equilibrium equations, expressions for the various receptor species can be derived and substituted into Equation (2). When this is done, and with conversion of all equilibrium association constants to equilibrium dissociation constants (i.e., $K_1 = 1/K_1'$ through $K_5 = 1/K_5'$), one obtains the general Langmuir binding expression for radioactive CD4 to CXCR4 with gp120 as a cofactor:

$$\wp = \frac{\left(\frac{[\text{CD}]}{K_1}\right)\left(\frac{[\text{gp}]}{K_2}\right) + \frac{[\text{CD}]}{K_5}}{\left(\frac{[\text{CD}]}{K_1}\right)\left(\frac{[\text{gp}]}{K_2}\right) + \frac{[\text{CD}]}{K_5} + \frac{[\text{gp}]}{K_3} + \frac{[\text{B}]}{K_4} + 1} \tag{3}$$

The observed affinity of the radiolabeled CD4 is given by the expression:

$$K_{\text{obs}} = \frac{K_1([\text{gp}]/K_3 + [\text{B}]/K_4 + 1)}{[\text{gp}]/K_2 + K_1/K_5} \tag{4}$$

The relevance of the above equations to SPA screening for ligands that interfere with CD4/gp120/receptor triad formation can be seen from the equation defining the IC_{50} of a nonradioactive ligand inhibitor (defined as the molar concentation of ligand [B] that blocks the radioactive binding signal by 50%). This can be obtained by solving Equation (4) for [B] = 0 and variable [B] for the equation yielding the concentration of [B] producing 50% inhibition of radioactive CD4 binding. This yields:

$$IC_{50} = K_4\left(\frac{[\text{CD}]}{K_1}\left(\frac{[\text{gp}]}{K_2} + \frac{K_1}{K_5}\right) + \frac{[\text{gp}]}{K_3} + 1\right) \tag{5}$$

From Equation (5) it can be seen that the Cheng Prusoff [2] equivalent expression for correction of the IC_{50} to the chemical equilibrium dissociation constant for the ligand/receptor complex (the system-independent measure of affinity) is given by:

$$K_4 = \frac{IC_{50}}{\left(\frac{[\text{CD}]}{K_1}\left(\frac{[\text{gp}]}{K_2} + \frac{[K_1]}{[K_5]}\right) + \frac{[\text{gp}]}{K_3} + 1\right)} \tag{6}$$

The desired endpoint of this assay is to define K_4, the molecular parameter defining the affinity of ligand [B] for the receptor, not the IC_{50}, which changes with assay conditions. Thus, the question is, what do variations in [gp] do to the relationship between K_4 and IC_{50}? At this point it is useful to define two ratios. The first is the ratio of the differential affinity of the gp120/CD4 complex vs. the affinity of gp120 for the receptor alone. This will be defined as $\phi = K_3/K_2$. Large values of ϕ mean that the preformed complex gp120/CD4 is the principal binding species to the receptor and that the affinity of gp120 for the receptor is relatively unimportant.

The second ratio to be defined is the differential affinity of CD4 for gp120 over the receptor. This is defined as $\psi = K_5/K_1$. Therefore, high values of ψ indicate that CD4 prefers to form the CD4/gp120 complex over binding to the receptor (i.e., the receptor functions as a true coreceptor for the formation of the CD4/gp120 complex). It should be noted that the known physiology of HIV entry into cells via this mechanism is consistent with high values for both ϕ and ψ. Equation (6) then becomes:

$$K_4 = \frac{IC_{50}}{\left(\frac{[CD]}{K_1}\left(\frac{[gp]}{K_2}+\frac{1}{\Psi}\right)+\frac{[gp]}{\Phi K_3}+1\right)} \tag{7}$$

It can be seen from Equation (7) that high values of ϕ and ψ (consistent with the known physiology), lead to a simplified Cheng-Prusoff relation of the form:

$$K_4 = \frac{IC_{50}}{\left(\frac{[CD]}{K_1}\right)\left(\frac{[gp]}{K_2}\right)+1} \tag{8}$$

From Equation (8) it can be seen that unknown variations in gp120 levels can lead to differences between the correction factor for the experimentally observed IC_{50} and the sought-after quantity K_4. However, this variation can be diminished if the minimal control signal is used for screening (i.e., minimal concentration of CD4 used to gain an acceptable signal-to-noise ratio). Therefore, as long as the magnitude of the original radioactive signal is kept constant by adjustment of radioactive CD4 levels, then variations in gp120 concentrations should have minimal effects on the screen. Equation (8) indicates how error can be reduced and the assay subsequently optimized.

Models can integrate many simple interactions, which, when connected, yield a highly complex potential behavior. One example is a radioligand binding to a receptor, which can then dimerize. This model is shown in Figure 1.2. Here it can be seen that the radioligand [A*] can bind to the receptor [R] to form a number of radioactive complexes that can be isolated, namely A*R, A*R–AR and A*R–A*R. Complications can arise when nonradioactive ligands bind to a receptor, which then dimerizes to an empty receptor site if there is an allosteric interaction between the two receptors. Usually, if an equilibrium is established between a concentration of radioligand [A*] and the receptor, and a nonradioactive ligand (denoted [A]) is added, then displacement of the radioligand will occur and a typical sigmoidal curve results. If dimerization occurs and there is an allosteric interaction between the receptors, then a complex situation, which may not be intuitively predictable, can result; here is where modeling can help define both the expected behavior and the optimal operating conditions for the system.

The conservation equation for the total receptor species is given as:

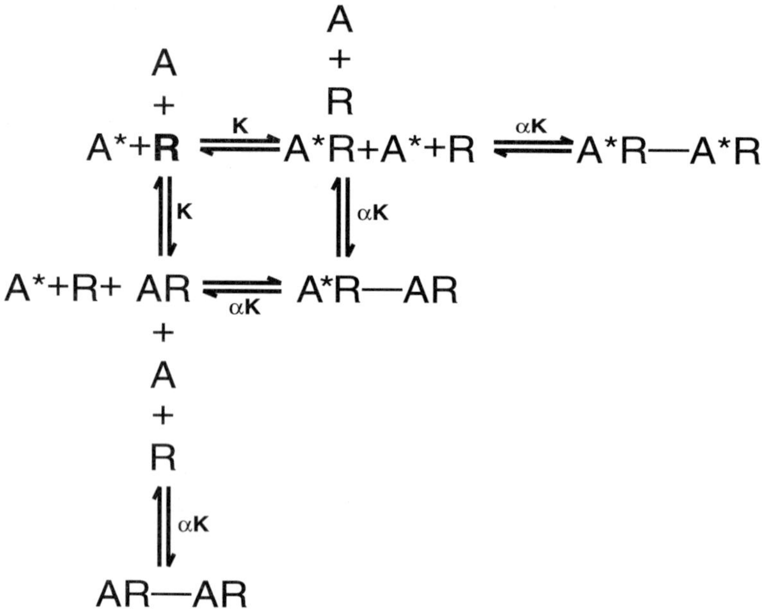

FIGURE 1.2 Receptor dimerization model. The binding of a radioligand ([A*]) or its nonradioactive counterpart ([A]) to a receptor promotes dimerization of the ligand-bound receptor to another receptor, which then can bind another ligand. The equilibrium association constant of the ligand for the receptor is denoted by K and the allosteric interaction coefficient (change in affinity between the ligand and receptor upon dimerization) is given by α. It is assumed that dimerization of nonligand bound receptor is neglible.

$$[R_{total}] = [R] + [AR] + [A^*R] + [A^*R\text{–}AR] + [AR\text{–}AR] + [A^*R\text{–}A^*R] \quad (9)$$

The radioactive signal (denoted $\wp$) from the binding assay emanates from species bound to [A*]:

$$\wp = \frac{[A^*R] + [A^*R - AR] + 2[A^*R - A^*R]}{[R_{total}]} \quad (10)$$

Using the equilibrium equations for the system, this equation becomes:

$$\wp = \frac{[A^*]K + \alpha[A^*][A]K^2 + 2\alpha[A^*]^2K^2}{1 + [A]K + [A^*]K + \alpha[A^*][A]K^2 + \alpha[A]^2K^2 + \alpha[A^*]^2K^2} \quad (11)$$

where K is the association constant. Displacement binding assays measure the effects of a nonradioactive ligand (in this case [A]) on a defined signal from a fixed concentration of radioactive ligand ([A*]); this signal (denoted ω) can be calculated from the ratio of Equation (11), with [A] = 0 and [A*] fixed, over the equations evaluated with [A*] fixed:

$$\omega = \frac{\left(1+\frac{\alpha[A^*]}{K_D}+\frac{2\alpha[A^*]}{K_D}\right)\left(1+\frac{[A^*]}{K_D}+\alpha\left(\frac{[A^*]}{K_D}\right)^2\right)}{\left(1+\frac{[A]}{K_D}+\frac{[A^*]}{K_D}+\frac{\alpha[A^*][A]}{K_D}+\alpha\left(\frac{[A]}{K_D}\right)^2\right)\left(1+\frac{2\alpha[A^*]}{K_D}\right)} \tag{12}$$

where $K_D = 1/K$. Using Equation (12), displacement curves for this system can be calculated. It is useful to examine two kinds of interaction between the two binding sites: 1.) the binding of one ligand is positively cooperative with the binding of the other ($\alpha > 1$) and 2.) where the binding is negatively cooperative ($\alpha < 1$). Figure 1.3A shows an interesting set of curves for a system where there is positive cooperativity (binding of one [A] and subsequent dimerization with another receptor increases the affinity for the second [A]). As can be seen from this figure, an apparently paradoxical *increase* in the radioactive signal is observed from addition of nonradioactive ligand if low concentations of radioligand are used, but this behavior disappears at higher concentrations of radioligand. If the cooperativity upon dimerization is negative (binding of [A] and subsequent dimerization leads to a decrease in the affinity of the receptor for the second [A]), then a different pattern of responses is observed. As shown in Figure 1.3B, monophasic displacement curves are seen at lower concentrations of radioligand followed by an increased binding of radioligand at higher concentrations. The modeling of this system would be a valuable prerequisite to its use if the increased radioactive binding was unexpected intuitively and thought to represent an experimental problem.

Another example of optimization of experimental systems can be found in the model for allosteric ligand interactions on monomeric receptors. The model, developed by Ehlert,[3] is shown in Figure 1.4A. It indicates that the affinity of a primary ligand (denoted [A*] as the tracer ligand because the quantity bound to the receptor can be measured) for the receptor is altered by a factor α (defined as the allosteric constant) when the receptor binds an allosteric ligand (denoted [B]). The cooperativity of binding can be positive ($\alpha > 1$) or negative ($\alpha < 1$). Thus, a given amount of tracer ligand is equilibrated with the receptor and then the effect of a range of concentrations of allosteric ligand added to determine the effect. An example of the effects of a negative allosteric ligand is shown in Figure 1.4B. It is useful to determine the relationship between the experimentally observed parameters of maximal ligand binding inhibition and potency of the negative allosteric ligand (determined as the IC_{50}, the concentration of allosteric ligand that inhibits the tracer signal by 50%) to the molecular parameters of α and K_I (the equilibrium dissociation constant of the allosteric ligand–receptor complex). The magnitude of these theoretical parameters is desired since they are system-independent and characterize the receptor and ligand in molecular terms. Modeling the system can establish this relationship as well as define the optimal conditions for determining them.

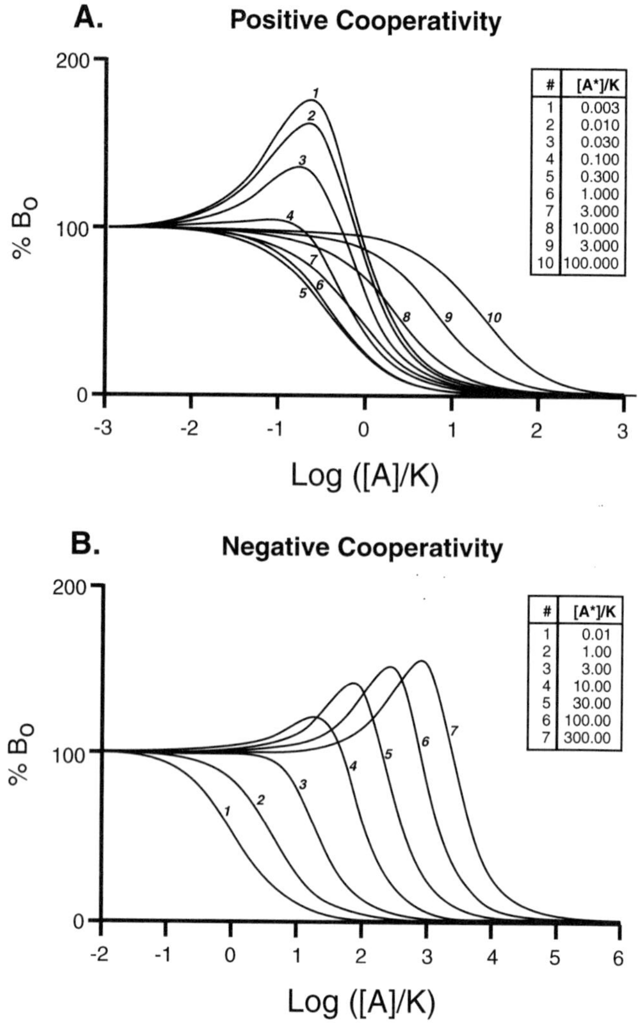

FIGURE 1.3 Effects of nonradioactive ligand ([A]) on signal from prebound radioactive ligand ([A*]) — see Figure 1.2 for model. Ordinates are a percentage of the signal obtained in the absence of nonradioactive ligand. Abscissae: logarithm of the molar concentration of nonradioactive ligand as multiples of the equilibrium dissociation constant of the ligand-receptor complex. A. Cooperativity of binding second ligand upon dimerization is positive. Numbers refer to curves calculated for various concentrations of radioligand. B. Cooperativity upon dimerization is negative.

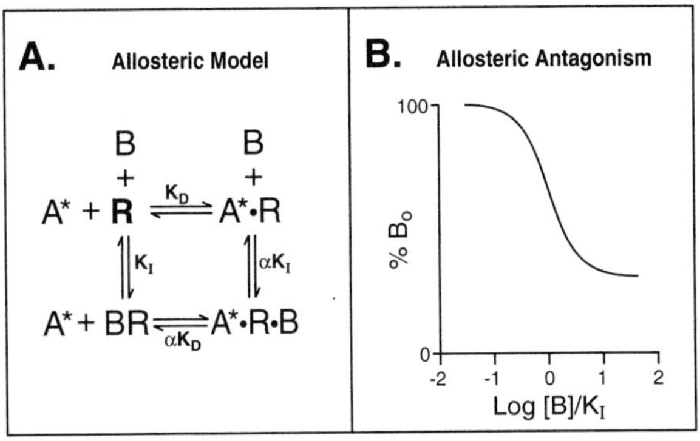

FIGURE 1.4 Allosteric model of independent binding of two ligands ([A] and [B]) to one receptor — after Ehlert.[3] A. Schematic illustration of the model. The equilibrium dissociation constant for binding of A and B to R is denoted as K_D and K_I, respectively. The change in affinity of either ligand after one of the ligands is bound given by the multiplication factor α. B. Effects of an allosteric antagonist ($\alpha < 1$) on the signal from a given prebound concentration of radioligand [A].

The signal from a radioligand [A], in the presence of a given concentration of allosteric antagonist [B], is given by:

$$\wp = \frac{\frac{[A]}{K_D}\left(1+\frac{\alpha[B]}{K_I}\right)}{\frac{[A]}{K_D}\left(1+\frac{\alpha[B]}{K_i}\right)+\frac{[B]}{K_I}+1} \tag{13}$$

The amount of radioligand bound before addition of the allosteric ligand is obtained from Equation (13) by setting [B] = 0 while the maximal effect of the allosteric ligand on the signal for the radioligand is obtained by setting $[B] \rightarrow \infty$. This defines the maximal scale of inhibition and, as shown in Figure 1.5A, is an approximation of the allosteric constant α.

After subtraction of the nonspecific binding, the midway point between B_o (initial radioligand binding whereby [B] = 0) and B_{min}, the minimum radioligand binding observed in the presence of saturating concentrations of allosteric antagonist (where allosterism is maximal and $[B] \rightarrow \infty$), can be calculated by subtracting the equations for B_o and B_{min}, dividing by 2, and adding B_{min}. When these values $(((B_o - B_{min})/2) + B_{min})$ and Equation (13) are equated, the following expression is obtained for the concentration of allosteric antagonist producing 50% inhibition (defined as the IC_{50} — see Figure 1.5B):

$$IC_{50} = \frac{K_I\left(1+\frac{[A]}{K_D}(1-\alpha)-\alpha\left(1+\frac{[A]}{K_D}\right)^2\right)}{1+\frac{[A]}{K_D}-\alpha\left(\frac{[A]}{K_D}(1-\alpha)-1\right)} \tag{14}$$

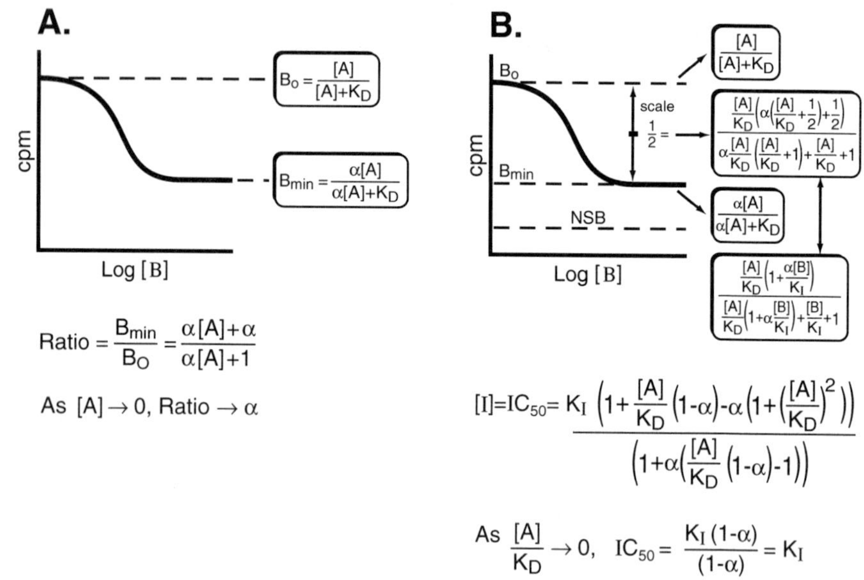

FIGURE 1.5 Meaning of asymptote (A) and IC_{50} (B) of an allosteric antagonist curve in terms of the Ehlert model.

It can be seen from this equation that if the concentration of radioligand is low (i.e., if $[A] << K_D$) then the IC_{50} will be equal to the K_I (see also ref. 3). This defines an optimal operating condition for an assay designed to detect allosteric antagonists and a relationship between an experimentally determined parameter (IC_{50}) and a molecular one (K_I).

1.3 MODELS WITH INTERNAL CHECKS

The most useful models are those which have, built within them, internal checks that allow experimenters to determine whether the system they are working with adheres to the rules of the model. One of the most famous checks on a pharmacological model can be found in Schild Analysis.

As defined by the Gaddum equation[4] for simple competitive antagonism, the receptor occupancy by an agonist ($\wp'_A$) is determined by the equilibrium dissociation constant of the agonist–receptor complex (K_A), the agonist concentration [A′], and the ratio of the antagonist concentration [B] divided by its equilibrium dissociation constant (K_B):[4]

$$\wp'A = \frac{\frac{[A']}{K_A}}{\frac{[A']}{K_A}+\frac{[B]}{K_B}+1} \tag{15}$$

In the absence of the antagonist, an agonist receptor occupancy of equal magnitude (denoted $\wp_A$) would be obtained at a lower concentration of agonist (denoted [A]) since $[B]/K_B = 0$; $\wp_A$ is therefore given by Equation (15) with $[B]/K_B = 0$. If it is assumed that equal receptor responses to agonists are obtained at equal receptor occupancies ($\wp_A = \wp'_A$), then the Schild equation for the ratio of equiactive doses of agonist ([A′]/[A]), which will be denoted by the term dr (dose-ratio = dr = [A′]/[A]) yields:

$$\frac{[B]}{K_B} = dr - 1 \tag{16}$$

This is used as its logarithmic metameter as the well-known Schild equation:[5]

$$Log(dr - 1) = Log[B] - LogK_B \tag{17}$$

Thus, a regression of the various dose-ratios obtained (as log (dr – 1)) upon the logarithms of the various antagonist concentrations that produced them (log [B]) should yield a straight line having a slope of unity. Therein lies the internal check. If the simple model used by Gaddum is operative in the experimental system (namely one agonist and one antagonist molecule vying for occupancy of one receptor), then the experimental Schild regression should be linear and have a slope of unity. Nonadherence to this internal rule can be useful data as well since it may be the herald of nonequilibrium steady-states within the system or receptor heterogeneity.[6-8]

Internal checks are even more essential for complex models since the number of parameters and reactants can be large and their interactions veiled in systems of compensation and averaged responses. For example, the extended ternary complex model for G protein-coupled receptors[9] (see Figure 1.6) leads to certain predictions between two experimentally observable responses. When two such independent responses can be determined, this can lead to an excellent opportunity for an internal

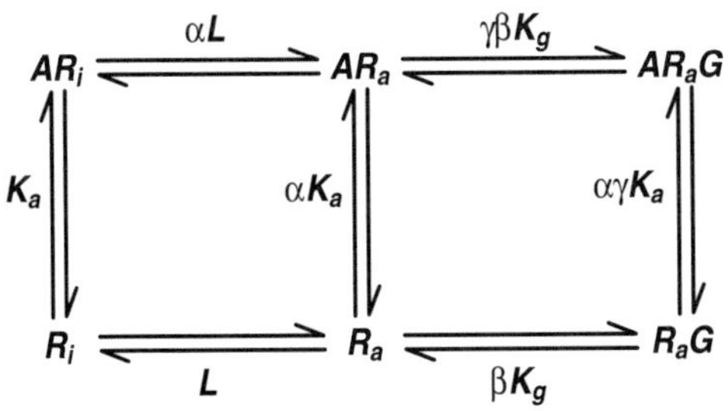

FIGURE 1.6 Extended ternary complex model for G protein-coupled receptors as defined by Samama et al.[9] Receptors coexist in an inactive (R_i) and active (R_a) state according to an equilibrium constant L. Ligand [A] can bind to either state, but only the active state (either as R_a or AR_a) can bind to G protein [G]. K_g and K_a denote the G protein and ligand equilibrium association constants, respectively.

check since the relationship between these responses can be defined, and then adherence (or nonadherence) to this relationship can be tested. Specifically, this model defines an equilibrium between receptors in the inactive state (R_i) and active state (R_a) and further dictates that receptors in the active state couple to G proteins to initiate GTP–GDP exchange and subsequent physiological response. One prediction of this model is that, as the receptor concentration increases (increased receptor expression in a recombinant system), so, too, will the constitutive ligand-independent response (due to spontaneous formation of the active receptor state). This is an experimentally observed phenomenon and has been documented for numerous receptor systems[9,10] (see ref. 11 for review). This effect can be modeled from equations describing the extended ternary complex model.

For this model, the conservation equation for G protein is:

$$[G_{tot}] = [G] + [R_aG] + [AR_aG] \tag{18}$$

The equation for the production of response ($\wp$), either constitutively or through an agonist ligand, [A], is:

$$\wp = \frac{\beta L \frac{[Ri]}{K_G}\left(1 + \alpha\gamma \frac{[A]}{K_A}\right)}{\frac{[A]}{K_A}\left(\alpha\gamma\beta L \frac{[Ri]}{K_G}\right) + 1 + \left(\beta L \frac{[Ri]}{K_G}\right)} \tag{19}$$

where $K_G = 1/K_g$ and $K_A = 1/K_a$. The basal constitutive receptor activity is given by Equation (19) when [A] = 0:

$$\text{Basal} = \frac{\beta L \frac{[Ri]}{K_G}}{1 + \beta L \frac{[Ri]}{K_G}} \tag{20}$$

Similarly, the potency (K_{obs}) of an agonist (as given by the concentration of agonist producing the half-maximal response) can be observed experimentally. The expression for this agonist potency can be derived from Equation (19) as:

$$K_{obs} = \frac{K_A\left(1 + \beta L \frac{[Ri]}{K_G}\right)}{\alpha\gamma\beta L \frac{[Ri]}{K_G}} \tag{21}$$

Equating $[Ri]/K_G$ from Equations (20) and (21) leads to the following relationship between observed constitutive activity and corresponding agonist potency in G protein-coupled receptors adhering to this model:

$$\frac{1}{K_{obs}} = \frac{\alpha\gamma(\text{Basal})}{K_A} \tag{22}$$

This inverse relationship predicts that with increased constitutive receptor activity comes an increased potency to agonists (decrease in the equilibrium dissociation constant of the agonist–receptor complex). Both constitutive receptor activity and agonist potency are experimentally observable behaviors, thus providing an internal check of the model for any given experimental system. Figure 1.7A shows the relationship between the constitutive activity obtained by transfection of melanophores with cDNA encoding human calcitonin receptors and the potency of human calcitonin in the same system. These data show that as constitutive activity (basal response) increases, the potency of human calcitonin also increases, suggesting agreement of this system with this model (Figure 1.7B).

Ultimately, a model becomes a tool for the prediction of endpoint responses from complex systems of numerous interactants. With such systems, it is often difficult, if not impossible, to intuitively determine the effect of changing one of the reactants or one of the interactions. Figure 1.8 shows a statistically and thermodynamically more complete model than the extended ternary complex model, shown in Figure 1.6, for a synoptic G protein-coupled receptor system.[12,13] As can be seen from the number of parameters and possible interactions, this model enters the realm of being heuristic (see Table 1.1). However, the model can still be a very useful predictor of complex behaviors. The following four examples show how the model can be used to predict effects not intuitively obvious.

It is well known that some receptors can be promiscuous with respect to the number of G proteins with which they can interact. For example, there is evidence

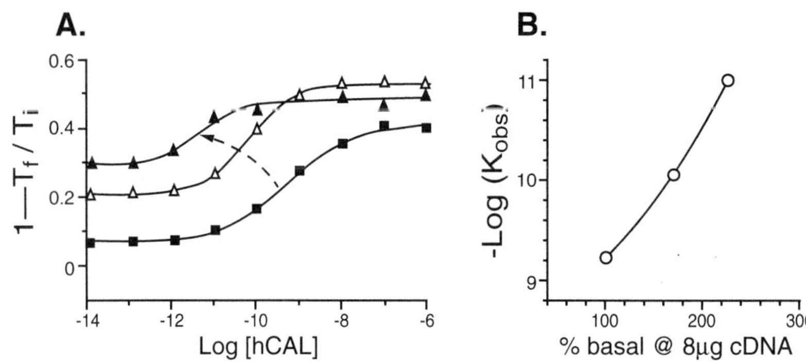

FIGURE 1.7 The effects of receptor concentration (human calcitonin receptor type 2, hCTR2) on basal constitutive response and sensitivity to human calcitonin. A. Dose-response curves to human calcitonin in melanophores transfected with cDNA for hCTR2 8 μg (filled squares), 16 μg (open triangles), and 32 μg (filled triangles). Ordinate values represent changes in light transmittance through a monolayer of melanophores. B. Correlation between basal constitutive activity (calculated as a percentage of that observed for transfection with 8 μg cDNA) and pEC_{50} (–log of the EC_{50}) for calcitonin.

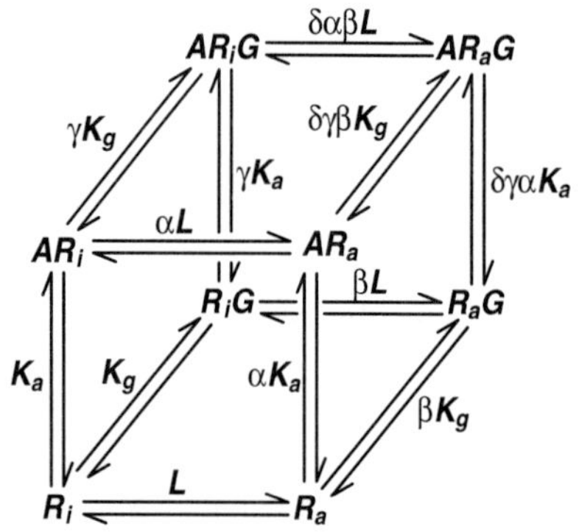

FIGURE 1.8 Cubic ternary complex model for G protein-coupled receptors (from Weiss et al.[12,13]). Similar to the extended ternary complex model shown in Figure 1.6 except the inactive state of the receptor (both ligand-bound and unbound) is allowed to interact with G protein.

that the β_2-adrenoceptor can activate G_s to, in turn, activate adenylate cyclase to cause increased cytosolic cyclic AMP concentrations. Concomitantly (for some agonists), this receptor also activates G_i to inhibit adenylate cyclase and thus reduce cytosolic cyclic AMP concentrations.[14-16] This system is shown schematically in terms of the cubic ternary complex model in Figure 1.9A. Intuitively, it might be supposed that the two opposing stimuli will result in a biphasic dose–response curve to β-adrenoceptor agonists in systems where this type of interaction could occur (Figure 1.9B). However, by modeling the system at several different stoichiometric setpoints (varying quantities of receptor and G protein as might be encountered in different recombinant or natural systems), an interesting phenomenon can be observed. Specifically, if the concentration of receptor is limiting (with respect to G protein), then *no* biphasic curve for cyclic AMP formation should be expected. For example, Figure 1.9C, shows concentration–response curves to a β-adrenoceptor agonist in a system with three different concentrations of receptor in cells containing 100 and 70 equivalents of G_s and G_i, respectively. Note that at an equivalent concentration of [R] = 30, the curve is monophasic. An increase in the receptor concentration to [R] = 100 leads to an increased maximal asymptote, but still no biphasic character to the curve. Only at receptor concentrations exceeding the G protein equivalent ([R] = 200 shown in Figure 1.9C) does the biphasic nature of the interaction become apparent. Thus, the model predicts that the β-adrenoceptor could be expressed in some cell lines and not demonstrate the biphasic G_i/G_s behavior while in others biphasic curves would be observed.

Another surprising prediction of the cubic ternary complex model emerges from exploration of model behavior; specifically, the behavior of G protein-coupled receptor systems toward agonists demonstrating "protean" efficacy. If an agonist produces an active receptor state that is less efficacious, in terms of coupling to G proteins, than the spontaneously formed constitutive active state ([Ra]), then under conditions of constitutive receptor activity, the agonist may actually inhibit constitutive activity (receptor changes from a highly efficacious active state to a lower efficacy active state). However, under conditions of no constitutive receptor activity (predominant

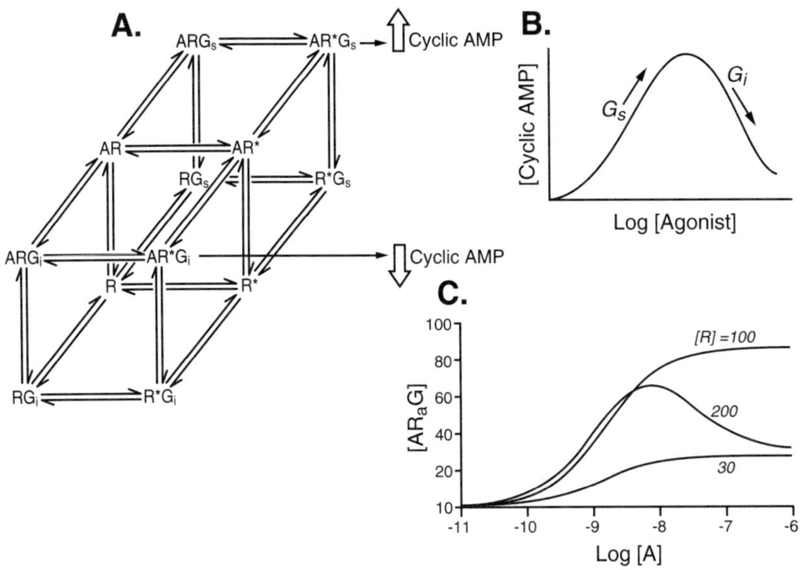

FIGURE 1.9 Cubic ternary complex model for G protein-coupled receptors showing the β-adrenoceptor ([R]) mediated activation of Gs-protein (causing elevation of cyclic AMP) and Gi-protein activation (which decreases cyclic AMP production). A. Schematic of the model. B. Expected response to a β-adrenoceptor agonist producing the opposing responses. C. Simulation with a receptor system having parameters [Gs] = 100, [Gi] = 70 (β_{Gs} = 30, β_{Gi} = 3, L = 0.1, K_{Gs} = 30, K_{Gi} = 10) for an agonist of α = 100, γ_{Gs} = 10, γ_{Gi} = 2, δ = 1, $K_A = 10^6$ (parameters for model shown in Figure 1.8). Panel C shows the differing response to the agonist at three different receptor concentrations.

receptor species = [Ri]), production of a lower efficacy active state results in positive agonism. Thus, depending on the setpoint of the receptor system, the ligand could function as a positive or inverse agonist (referred to as "protean agonism" after the Greek god, Proteus[17,18]). Protean agonism has been observed for the β-adrenoceptor agonist dichloroisoproterenol (DCI).[19] In a constitutively active β-adrenoceptor system, DCI produced positive agonism. However, after desensitization of the receptor by prior exposure to the powerful β-adrenoceptor agonist isoproterenol, DCI produced inverse agonism. These data showed that, with a change in the condition of the receptor system, the efficacy of this ligand changed from positive to negative. The molecular mechanism for such a change is not intuitively obvious but can be mimicked with the cubic ternary complex model by assuming that DCI promotes an active state of the receptor that is a less efficacious active state than the constitutively active one. The desensitization step can be modeled by reducing the affinity of the receptor for G protein (reduction in the magnitude of β). Intuitively, it would be expected that the desensitization step (reduction in the affinity of the active receptor for G protein) would necessarily decrease the constitutive activity of the system. There are a wide range of conditions under which this should occur — an example is shown in Figure 1.10A. However, the experimental system showed no such decrease in the constitutive activity, only a reversal of agonism for DCI.

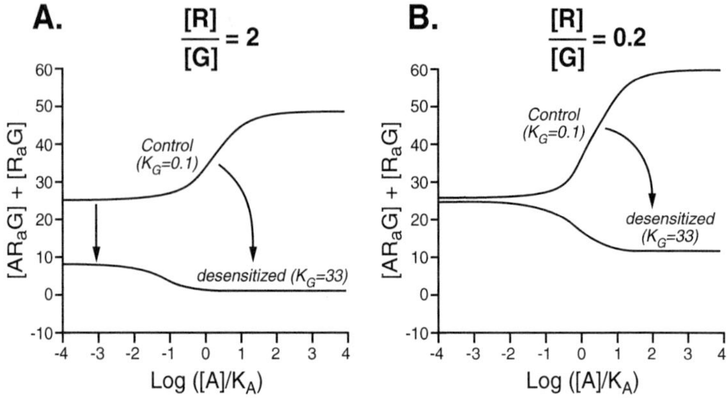

FIGURE 1.10 Effects of a protean agonist in a defined control and desensitized receptor system. Simulation with a receptor system having parameters $\beta = 10$, L = 0.1, [R] = 100 for an agonist of $\alpha = 300$, $\gamma = 0.003$, $\delta = 1$, $K_A = 10^6$ (parameters for model shown in Figure 1.8). Desensitization was modeled by decreasing the affinity of the receptor for G protein from $K_G = 30$ to $K_G = 0.003$. Panel A shows effect of desensitization on basal and agonist response in a system with relatively low levels of G protein ([G] = 50) and panel B, the lack of effect on basal response with higher concentrations of G protein in the system ([G] = 500).

Therefore, the model appeared to be at odds with the experimental finding. By systematically changing the system parameters (receptor/G protein relative stoichiometry), a surprising observation can be made. Specifically, if the concentration of G protein is substantially greater than the concentration of receptor, then a "buffering" effect of the desensitization is produced that allows the constitutive activity to be maintained in the presence of receptor desensitization while still allowing the reversal of DCI activity (Figure 1.10B). Thus, by exploring the behavior of the complex model beyond intuitive limits, a conclusion was reached that mated the model to the data in a positive fashion.

Models can also be useful in determining the impact of random error or variability on complicated systems. While the effects of varying a single element in a system are usually easily determined, the impact of many variations on multiple elements may not. For example, the cubic ternary complex model for G protein-coupled receptors (Figure 1.8), predicts that the coupling of receptors to G proteins should produce a response, the magnitude of which depends on system factors, namely the concentration of available G protein for complexation and the relative spontaneous activity of the receptor (which can be affected by an ionic environment[20]). Thus, as cDNA for a given receptor is transfected into different host cells, the ability of agonists to couple to G proteins can vary, and thus, so will the response to the agonist. One way to estimate how a collection of apparently small changes (e.g., 15%) in host cell factors can affect the totality of the observed agonist activity is to impose a random error function on the system factors and repeatedly calculate the end response (in this case the magnitude of the agonist response). Figure 1.11A shows a schematic of the process for the cubic ternary complex model and Figure 1.11B

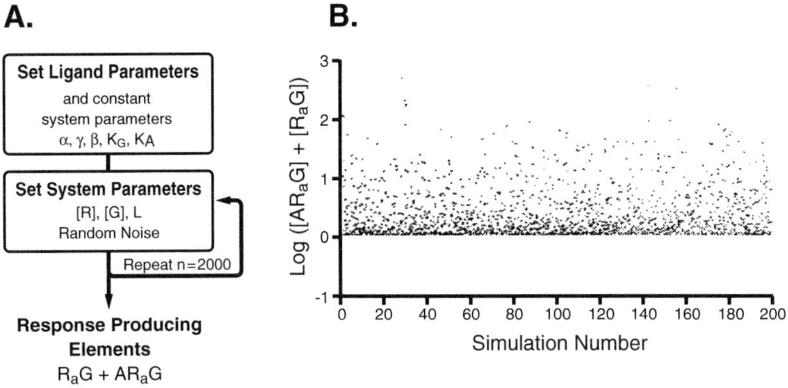

FIGURE 1.11 Repeated simulations of response producing species [AraG] and [RaG] for an agonist (α = 10, γ = 10). A. Schematic of iteration procedure. The systems parameters K_G and β and agonist parameters α and γ were kept constant. The system parameters [R], [G], and L were defined, then subjected to 15% random error and the levels of [RaG] and [AraG] calculated. This latter procedure was automated and repeated 2000 times. B. The resulting levels of [RaG] and [AraG] (response) are shown on a logarithmic scale as dots as a function of the simulation number.

the result of 2000 random variations in the concentration of G protein and allosteric constant L. As can be seen from this simulation, considerable variation in the response to the agonist can be produced by transfection of the receptor into different cells. This is not measurement error since each estimation of response is the "correct" one for that synoptic system, but rather it is biological variation that would be expected in recombinant systems. The simulation is useful in developing a sense for the magnitude of small changes when summed in a complex system.

Systematic exploration of a complex model can lead to general insights. For example, receptor theory predicts that the relative maximal responses of two agonists, providing they produce maximal responses less than the system maximum (i.e., they are partial agonists), is solely dependent upon the efficacy of the agonists and is independent of affinity. Therefore, in theory, the relative maxima of two agonists, in an appropriate system, could be used to quantify their ligand properties of relative intrinsic efficacy and thus characterize them on a molecular scale.

This is predicted by the cubic ternary complex model as well. In the model, the response to an agonist is given as:[21]

$$\wp = \frac{\beta L \frac{[G]}{K_G}\left(1 + \delta\alpha\gamma\frac{[A]}{K_A}\right)}{\frac{[A]}{K_A}\left(1 + \alpha L + \gamma\frac{[G]}{K_G}(1 + \delta\alpha\beta L)\right) + \frac{[G]}{K_G}(1 + \beta L) + 1 + L} \tag{23}$$

The maximal response (when $[A] \rightarrow \infty$.) from Equation (23) is:

$$E_{max} = \frac{\delta\alpha\beta\gamma L \frac{[G]}{K_G}}{1 + \alpha L + \gamma \frac{[G]}{K_G}(1 + \delta\alpha\beta L)} \tag{24}$$

The relative maxima of two agonists (denoted [A] and [B] as subscripts) is given as:

$$\text{Rel. Max} = \frac{(\delta_A \alpha_A \gamma_A)\left(1 + \alpha_B L + \gamma_B \frac{[G]}{K_G}(1 + \delta_B \alpha_B \beta L)\right)}{(\delta_B \alpha_B \gamma_B)\left(1 + \alpha_A L + \gamma_A \frac{[G]}{K_G}(1 + \delta_A \gamma_A \beta L)\right)} \tag{25}$$

It can be seen, that while receptor theory suggests that the relative maxima should be system independent (i.e., depend only on the ligand parameters δ, α, and γ), Equation (25) shows that the system variables [G] and L appear in the expression. Therefore, there are experimental conditions whereby the system does impose restraints on the estimation of efficacy by this method, and others where it does not. In this way a systematic exploration of the model behavior around these system variables is useful to define the types of systems where efficacy estimation by this method is suspect.

Figure 1.12 shows a three dimensional surface varying the system parameters L and [G] and calculating the relative maximal responses of two defined agonists. The contoured surfaces show the variation of the relative maxima with changing system parameters (where efficacy estimates are suspect). Also shown are the concentration response curves for the two agonists according to Equation (23). An interesting behavior of this system is revealed. The variation in the maxima (i.e., estimate of efficacy) begins when there is a basal constitutive agonist-independent response in the system, i.e., the dubious estimates of efficacy occur when the system is constitutively active. Since this latter response (constitutive activity) is observable experimentally, this defines a useful criterion for receptor systems designed to estimate the relative efficacy of agonists, namely an absence of constitutive activity.[22]

1.4 CONCLUSIONS

Quantitative mathematical models of drug receptor systems can be used to predict the drug effect in receptor systems, the strengths and weaknesses of such systems, and where the experimental conditions are optimal. The imposition of quantitative constraints defines rules of behavior for receptor systems that greatly enhance the ability of the researcher to assess the pharmacologic significance of experimental findings. From this standpoint, mathematical modeling is an essential part of the drug and drug receptor classification process.

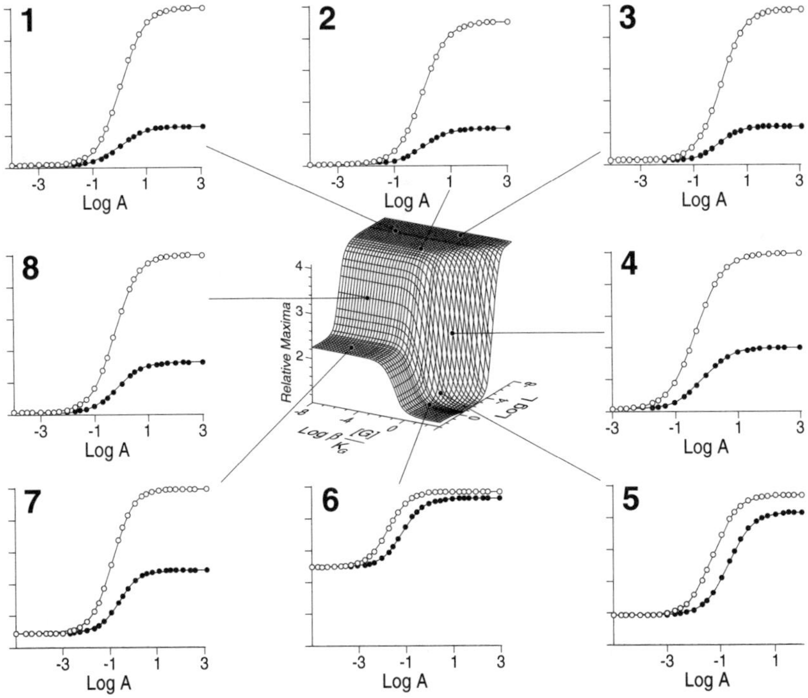

FIGURE 1.12 Effect of varying systems parameters L or $\beta[G]/K_G$ on the relative maximum responses to two agonists, one with $\alpha = \gamma = 5$ and the other $\alpha = \gamma = 10$. Also shown are the dose–response curves to the agonists for 8 different system setpoints. It can be seen that the relative maxima are relatively constant (surface points 8, 1, 2, 3, 4) when there is no elevated basal response, but begin to alter as the system becomes constitutively active (points 7, 6, 5). From Reference 22 with permission.

REFERENCES

1. Feng, Y., Broder, C. C., Kennedy, P. E., and Berger, E. A., HIV-1 entry cofactor: functional cDNA cloning of a seven-transmembrane, G protein-coupled receptor, *Science,* 272,872, 1996.
2. Cheng, Y. C. and Prusoff, W. H., Relationship between the inhibition constant (Ki) and the concentration of inhibitor which causes 50 percent inhibition (I_{50}) of an enzymatic reaction, *Biochem. Pharmacol.,* 22, 3099, 1973.
3. Ehlert, F. J., Estimation of the affinities of allosteric ligands using radioligand binding and pharmacological null methods, *Mol. Pharmacol.*, 33, 187, 1988.
4. Gaddum, J. H., The quantitative effects of antagonistic drugs, *J. Physiol. (Lond.)*, 89, 7P, 1937.
5. Arunlakshana, O. and Schild, H. O., Some quantitative uses of drug antagonists, *Br. J. Pharmacol.,* 14, 48, 1959.
6. Kenakin, T. P., The Schild regression in the process of receptor classification, *Can. J. Physiol. Pharmacol.,* 60, 249, 1982.

7. Kenakin, T. P., Tissue response as a functional discriminator of receptor heterogeneity: Effects of mixed receptor populations on Schild regressions, *Mol. Pharmacol.,* 41, 699, 1992.
8. Kenakin, T. P., Affinity, in *Pharmacologic Analysis of Drug-Receptor Interaction,* 3rd ed., T. P. Kenakin, Lippincott-Raven, New York, 1997, Chap 8.
9. Samama, P., Cotecchia, S., Costa, T., and Lefkowitz, R. J., A mutation-induced activated state of the β_2-adrenergic receptor: Extending the ternary complex model, *J. Biol. Chem.,* 268, 4623, 1993.
10. Van Sande, J., Swillens, S., Gerard, C., Allgeier, A., Massart, C., Vassart, G., and Dumont, J. E., In Chinese hamster ovary K1 cells dog and human thyrotropin receptors activate both cyclic AMP and the phosphatidylinositol 4,5-bisphosphates cascades in the presence of thyrotropin and the cyclic AMP cascade in its absence, *Eur. J. Biochem.,* 229, 338, 1995.
11. Kenakin, T. P., The classification of seven transmembrane receptors in recombinant expression systems, *Pharmacol. Rev.*, 48, 413, 1996.
12. Weiss, J. M., Morgan, P. H., Lutz, M. W., and Kenakin, T. P., The cubic ternary complex receptor-occupancy model. I. Model description, *J. Theoret. Biol.,* 178, 151, 1996.
13. Weiss, J. M., Morgan, P. H., Lutz, M. W., and Kenakin, T. P., The cubic ternary complex receptor-occupancy model. II. Understanding apparent affinity, *J. Theoret. Biol.,* 178, 169, 1996.
14. Chaudhry, A., MacKenzie, R. G., Georgic, L. M., and Granneman, J. G., Differential interaction of β_1- and β_3-adrenergic receptors with G_i in rat adipocytes, *Cellular Signalling,* 6, 457, 1994.
15. Xiao, R. P., Avdonin, P., Zhou, Y. Y., Cheng, H., Akhter, S. A., Eschenhagen, T., Lefkowitz, R. J., Kock, W. J., and Lakatta, E. G., Coupling of β_2-adrenoceptor to G_i proteins and its physiological relevance in murine cardiac myocytes, *Circ. Res.,* 84, 43, 1999.
16. Soeder, K. J., Snedden, S. K., Cao, W., Della Rocca, G. J., Daniels, K. W., Luttrell, L. M., and Collins, S., The β_3-adrenergic receptor activates mitogen-activated protein kinase in adipocytes through a G_i-dependent mechanism, *J. Biol. Chem.,* 274, 12017, 1999.
17. Kenakin, T. P., Pharmacologic Proteus?, *Trends Pharmacol. Sci.,* 16, 256, 1995.
18. Kenakin, T. P., Protean agonists: Keys to receptor active states?, *Ann. New York Acad. Sci.,* 812, 116, 1997.
19. Chidiac, P., Nouet, S., and Bouvier, M., Agonist-induced modulation of inverse agonist efficacy at the β_2-adrenergic receptor, *Mol. Pharmacol.*, 50, 662, 1996.
20. Costa, T. and Herz, A., Antagonists with negative intrinsic activity at δ-opioid receptors coupled to GTP-binding proteins, *Proc. Natl. Acad. Sci. U.S.A.,* 86, 7321, 1989.
21. Kenakin, T. P., The evolution of drug receptor models: The cubic ternary complex model for G protein-coupled receptors, *The Pharmacology of Functional, Biochemical, and Recombinant Receptor Systems, Handbook of Pharmacology*, vol. (in press), T. P. Kenakin and J. A. Angus, eds., Springer-Verlage, Heidelberg, Chapter 2.2.
22. Kenakin, T. P., Efficacy in drug-receptor theory: Outdated concept or under-valued tool?, *Trends Pharmacol. Sci.,* 20, 400, 1999.

2 The Ternary Complex Model

Frederick J. Ehlert

CONTENTS

0-8493-0100-9/01/$0.00+$.50

2.1 INTRODUCTION

G protein-linked receptors represent a huge family of proteins, and the most frequently used model for describing how drugs interact with these receptors is the "ternary complex model." This model, originally known as the "mobile receptor hypothesis," was derived by Jacobs and Cuatrecasas[1] to describe how the binding of an agonist to its receptor might cause the receptor to associate with another effector protein in the membrane, thus giving rise to a ternary complex of agonist, receptor, and effector. De Lean and coworkers[2] adapted this model to G protein-linked receptors in their analysis of the binding properties of agonists and antagonists to β-adrenergic receptors. In this context, the term "ternary complex" refers to the high affinity, stable composite of agonist, receptor, and G protein that can be measured in membrane binding assays in the absence of guanine nucleotides. However, in the intact cell during receptor activation, the relative abundance of the ternary complex is actually quite low, and it has both reduced affinity and stability. Under these conditions, it is replaced by a quaternary complex consisting of agonist, receptor, G protein, and guanine nucleotide. Regardless, the name "ternary complex" has stuck, and the model enjoys widespread use in receptor pharmacology.

Modern methods in cell biology and biophysics now give us a picture of receptor structure and function in exquisite detail. This picture contrasts with the rather concise mathematical language of receptor models. How can a simple scheme of an equilibrium between ground and active conformations of the receptor ($R \leftrightarrow R^*$), for example, adequately describe the complex repositioning of domains in the receptor during activation? Clearly, an equilibrium expression lacks this degree of visual resolution. Nevertheless, models do provide a means of measuring the activity of receptor structures, without which we would be unable to assess their significance, however exquisitely defined they may be. In other words, structural techniques tell us what the receptor looks like, whereas modeling methods enable us to measure its activity. Therefore, these two approaches are not directly comparable; rather, they are complementary and represent distinct components in our knowledge of receptors.

Affinity is an indispensable parameter in receptor modeling, and unfortunately, much confusion exists in the pharmacological literature because of inconsistent use of this term. Generally speaking, affinity is a measure of the "stickiness" of ligand–protein and protein–protein interactions, which can be represented quantita-

TABLE 2.1
Hierarchical Levels of Analysis for Receptor Models

Level 1: *Observed binding parameters:* The observed dissociation constants (K_{obs}) of the drug-receptor complex (DR) and the drug-receptor-G protein complex (*DRG*)

Level 2: *Microscopic constants of receptor complexes:* The microscopic affinity constants of the durg for the *DR* and *DRG* complexes (K_1 and αK_1) and the microscopic affinity constants of the G protein for the *RG* and *DRG* complexes (K_2 and αK_2).

Level 3: *Microscopic constants of receptor conformations:* The microscopic affinity constants of the drug for the ground (R_s) and active (R_s*) conformations of the receptor (K_a and aK_a) and the microscopic constants for the equilibrium between the ground and active conformations of different receptor complexes (K_c, aK_c, abK_c and bK_c for *R, DR, DRG* and *RG*, respectively).

tively by an equilibrium dissociation constant. However, in receptor modeling, there are three ways to define an equilibrium dissociation constant, which give rise to three hierarchical levels of explanation for the ternary complex model (see Table 2.1). On the surface, we have the *observed dissociation constant.* This parameter or its reciprocal (i.e., *observed affinity constant*) defines the relationship between the ligand concentration and occupancy of a saturable component of binding. There may be one or more components of binding, each distinguishable by an observed dissociation constant (i.e., the concentration of ligand required for half-maximal occupation of the receptor component). The second level of complexity addresses the different types of receptor complexes, like the ternary complex (i.e., agonist-receptor-G protein) and binary agonist-receptor complex. Under some conditions of the ternary complex model, binding is consistent with a simple one-site model, and the observed dissociation constants of the ternary and agonist-receptor complexes are the same. However, let's imagine that we could affix the receptor to the G protein so that the two cannot dissociate. If we could measure the dissociation constant of the agonist for this complex, we would find that the agonist has much higher affinity. This measure of affinity is called the *microscopic affinity constant of the agonist for the receptor-G protein complex.* The final layer of complexity addresses the different shapes of the receptor. It is an implicit assumption in the ternary complex model that the agonist causes the receptor to associate with the G protein by changing the shape of the receptor and enabling complementary domains on the receptor and G protein to stick together. If we could lock the receptor into the shape that binds the G protein, we would find that the measure of agonist affinity for this active state is greater than its microscopic affinity constant for the receptor-G protein complex or any observable component of binding. The parameter that defines the affinity of a ligand for a discrete conformation of the receptor is called the *microscopic affinity constant of the particular state of the receptor* (e.g., active state). In this chapter, the ternary complex model will be explained at these three different levels. Others have written that the simple ternary complex model is invalid because it does not address the ground and active conformations of the receptor. My intent is to show that the different hierarchical levels of explanation are all internally consistent, and thus, they are all valid. The situation is akin to a physicist's explanation of matter in terms of atoms on the one hand and in terms of fundamental particles on the other. Both explanations are valid and may be equally satisfying depending upon our level of focus.

In this chapter, the details of the ternary complex model and its relationship to classic receptor theory will be described. The behavior of the model in both the presence and absence of guanine nucleotides will also be described. Special attention will be given to comparing the activity of native receptors and their recombinant versions expressed in cell lines. Finally, the topic of comparing the activity of different receptor mutants will be addressed. My intent is to take a broad view of receptor modeling and to show how the ternary complex model is most often consistent with the mathematics of classic receptor theory.

2.2 SIMPLE TERNARY COMPLEX MODEL

In its original form, the ternary complex model describes the behavior of G protein-linked receptors in the absence of guanine nucleotide. These conditions are met in experiments on broken cell homogenates in which guanine nucleotides have been washed away; however, they are far removed from those of the intact cell where GTP is present in near millimolar concentrations.[3-5] Nevertheless, important clues about the mechanism of G protein activation are easily revealed without the influence of guanine nucleotides, making the simple ternary complex model highly significant.

The simple ternary complex model is representative of an analysis at the second hierarchical level shown in Table 2.1. The model is based on the assumption that the receptor (R) and G protein (G) diffuse within the plane of the membrane and bind with each other as governed by the microscopic affinity constant, K_2. The drug (D) has an allosteric effect on this equilibrium by binding to its recognition site on the receptor as shown below:

$$
\begin{array}{ccccc}
 & & K_1 & & \\
D + R + G & & \longleftrightarrow & & DR + G \\
K_2 & \updownarrow & & \updownarrow & \alpha K_2 \\
D + RG & & \longleftrightarrow & & DRG \\
 & & \alpha K_1 & &
\end{array}
\tag{1}
$$

In this scheme, K_1 denotes the microscopic affinity constant of the drug for the receptor, DR denotes the drug-receptor complex, and RG denotes the receptor-G protein complex. The law of conservation of standard free energy requires that the drug and the G protein have reciprocal effects on each other.[6] Thus, the effect that D has on the binding of G is equivalent to the effect that G has on the binding of D. This mutual effect is known as heterotropic cooperativity, and it is denoted by α. If the drug enhances the binding of the G protein to the receptor, then there is positive cooperativity ($\alpha > 1$). Alternatively, if the drug allosterically inhibits the binding of the G protein to the receptor, there is negative cooperativity ($0 < \alpha < 1$). Thus, αK_1 denotes the microscopic affinity constant of D for RG; αK_2 denotes the microscopic affinity constant of G for DR; and DRG denotes the resultant ternary complex. The solution to this square of equilibria is given in the Appendix of this chapter. To understand the model better, it is useful to introduce the ground (R_s) and active (R_s*) states of the receptor, which represent an analysis at the third hierarchical level (see Table 2.1).

2.2.1 Conformational Analysis of the Ternary Complex Model

The increase in the activity of the receptor upon agonist binding can be attributed to a change in the shape of the receptor.[7] Thus, there must be at least two states of the receptor, ground and active, and the degree of receptor activation is proportional to the amount of receptor in the active state (R_s*). The scheme describing the binding of a drug to a receptor existing in two conformational states is shown below:

$$\begin{array}{ccc} & aK_a & \\ D + R_s^* & \longleftrightarrow & DR_s^* \\ K_c \updownarrow & & \updownarrow aK_c \\ D + R_s & \longleftrightarrow & DR_s \\ & K_a & \end{array} \qquad (2)$$

In this scheme, K_a and aK_a denote the microscopic affinity constants of D for the ground and active states of the receptor, respectively. The equilibrium between conformations of the bound ($DR_s \leftrightarrow DR_s^*$) and unbound ($R_s \leftrightarrow R_s^*$) forms of the receptor are governed by the constants aK_c and K_c, respectively. Thermodynamic principles require that there is no net flux or cycling around this square of equilibria just as in the case of the ternary complex model. This constraint requires that $K_a aK_c = K_c aK_a$.[6] For a receptor having negligible constitutive activity, very little receptor is in the active state in the basal condition ($K_c << 1$). The process of agonist-induced conformational change can occur by two mechanisms, conformational induction and conformational selection.[7,8] In the former mechanism, the agonist binds to the ground state and then subsequently induces the active state:

$$D + R_s \leftrightarrow DR_s \leftrightarrow DR_s^* \qquad (3)$$

In conformational selection, the agonist binds selectively to the active state of the receptor:

$$D + R_s \leftrightarrow D + R_s^* \leftrightarrow DR_s^* \qquad (4)$$

The consequences of either mechanism are the same at equilibrium. Over a particular range of values, the ratio of affinity constants for the ground and active conformations (i.e., $aK_a/K_a = a$) is proportional to the agonist's ability to generate the active conformation. This ratio is also related to the intrinsic efficacy of the agonist as described in a later section. Antagonists are drugs that lack selectivity for the active state. This condition can arise if the antagonist has similar affinity for the ground and active conformations ($K_a \cong aK_a$, i.e., $a \cong 1$) or if the antagonist has selectivity for the ground state ($K_a > aK_a$, i.e., $a < 1$).[9] If the receptor lacks constitutive activity, there will be no observable difference in the behavior of the two types of antagonists in most functional assays. However, antagonists with selectivity for the ground state will inhibit the basal activity of constitutively active receptors.[8,10] These agents are commonly known as "inverse agonists." However, since these agents generate the ground state of the receptor, they should be recognized by a term incorporating the word "antagonist." A more suitable alternative is the term "intrinsic antagonist," which recognizes the intrinsic capacity to generate the ground state of the receptor (intrinsic antagonistic activity).

It can be shown that scheme 2 is consistent with a simple one-site model having an affinity constant (K_1) equal to:

$$K_1 = (K_a + aK_aK_c)/(K_c + 1) \tag{5}$$

Scheme 2 actually represents an expanded form of the upper side of the ternary complex model shown in scheme 1 (i.e., $D + R \leftrightarrow DR$). Each receptor term (e.g., R) in the simple ternary complex model is composed of two components (R_s and R_s*) in scheme 2, with each component or state designated with a subscript "s" to distinguish it from terms used in scheme 1. We can also expand the other sides of the simple ternary complex model to yield the cube of equilibria shown in Figure 2.1. Each vertical face of the expanded ternary complex model corresponds to one side of the simple ternary complex model. Also, each microscopic affinity constant of the simple ternary complex model can be expressed in terms of the microscopic constants of the different states of the receptor as shown above for K_1 in Equation (5). The other affinity constants in the simple ternary complex model are given by:

$$K_2 = (K_b + bK_bK_c)/(K_c + 1) \tag{6}$$

$$\alpha K_1 = (K_a + abK_aK_c)/(bK_c + 1) \tag{7}$$

$$\alpha K_2 = (K_b + abK_bK_c)/(aK_c + 1) \tag{8}$$

In these equations, K_b denotes the microscopic affinity constant of G for R_s, and bK_b denotes the microscopic affinity constant for the binding of G to R_s*. It is also possible to express the cooperativity (α) in terms of the microscopic constants of the ground and active states of the receptor:

$$\alpha = \frac{(K_a + abK_aK_c)(K_c + 1)}{(K_a + aK_aK_c)(bK_c + 1)} \tag{9}$$

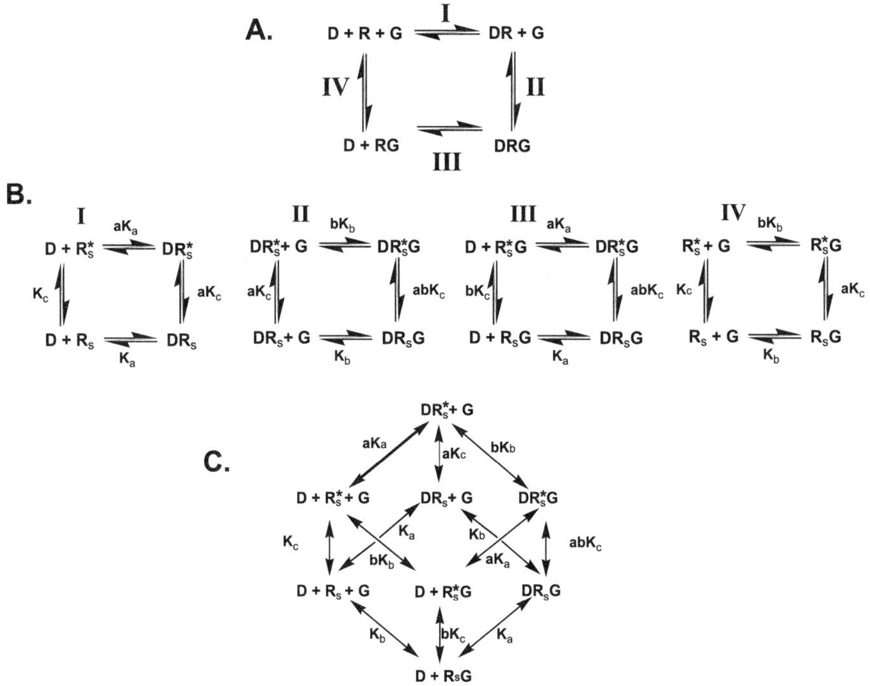

FIGURE 2.1 Conformational analysis of the ternary complex model. A. The simple ternary complex model. B. Each of the four equilibria in the simple ternary complex model has been isolated and expanded to show the ground (R_s) and active (R_s*) conformations of the receptor. C. The four expanded equilibria of the simple ternary complex model have been combined to form a cube of equilibria referred to as the expanded ternary complex model. This model is not the same as the "Cubic Ternary Complex Model" of Weiss et al.,[11] which is more complicated and has an undefined number of receptor conformations. The definitions of the variables and equilibrium constants are described in the text and Appendix in this chapter.

$$\alpha = \frac{(K_b + abK_bK_c)(K_c + 1)}{(K_b + bK_bK_c)(aK_c + 1)} \tag{10}$$

The solution to the expanded form of the ternary complex model shown in Figure 2.1 is given in the Appendix.

As shown in Figure 2.1, affinity is determined by the state of the receptor in the expanded ternary complex model. For example, on the bottom side of the model, the affinity of D for R_s (K_a) is the same as the affinity of D for R_sG. The affinities are the same because the shape of the receptor determines affinity, and the shape of the receptor is constant on the bottom side. The same concept applies to the top side of the model, where the affinity of D for R_s* and R_s*G is the same. However, in the simple ternary complex model, each type of receptor complex (i.e., DR and DRG) exhibits a different microscopic affinity constant. This situation arises because each complex actually represents an equilibrium between two states (R_s and R_s*) of differing affinities, and the equilibrium between the two states is different for each complex. Thus, the microscopic affinity constant of a receptor complex represents a kind of weighted average of the two states. In the expanded ternary complex model (Figure 2.1), ligand affinity changes on the vertical sides of the model, increasing for the agonist as the receptor changes from the ground state (bottom) to the active state (top).

The expanded ternary complex model shown in Figure 2.1 should not be confused with the cubic ternary complex model of Weiss et al.[11] These investigators proposed that the unbound receptor exists in two conformations, active (R_a) and inactive (R_i), each having a distinct microscopic affinity constant (αK_a and K_a, respectively). However, the affinity of the agonist for different complexes of the same state of the receptor are different in the cubic ternary complex model. That is, the two microscopic affinity constants for the inactive complexes (DR_i, DR_iG) are different as well as those for the two active complexes (DR_a and DR_aG), yielding a total of four microscopic affinity constants (i.e., K_a, γK_a, αK_a, and $\delta\alpha\gamma K_a$, respectively) for the agonist. This condition can only arise if there are more than two conformations of the receptor. In contrast, the expanded ternary complex model shown in Figure 2.1 only considers two states of the receptor, R_s and R_s*.

2.2.2 Behavior of the Ternary Complex Model with Excess G Protein

The behavior of the ternary complex model depends on the ratio of G protein to receptor (G_T/R_T; where G_T and R_T denote the total concentrations of G protein and receptor, respectively). When there is an abundance of G protein (i.e., $G_T/R_T >> 1$), then the binding of agonists and antagonists to the receptor is consistent with a one-site model having an observed dissociation constant (K_{obs}) equal to:

$$K_{obs} = \frac{1 + K_2 G}{K_1 + \alpha K_1 K_2 G} \qquad (11)$$

In this equation, K_{obs} is expressed in terms of the microscopic affinity constants of the simple ternary complex model. By substituting Equations (5) through (8) for these constants, we can express the K_{obs} in terms of the microscopic constants of the expanded ternary complex model:

$$K_{obs} = \frac{1 + K_c + GK_b + GbK_bK_c}{K_a + aK_aK_c + GK_aK_b + GabK_aK_bK_c} \qquad (12)$$

Both expressions [i.e., Equations (11) and (12)] are equivalent; they simply represent different levels of analysis. Therefore, the simple ternary complex model can accommodate any outcome of the expanded ternary complex model (e.g., constitutive receptor activity). The expanded ternary complex model defines the parameters of the simple ternary complex model in terms of more fundamental parameters. Moreover, for each receptor complex in the simple ternary complex model, the expanded ternary complex model shows what proportion of the complex is in the ground or active states (see Tables 2.2 through 2.5).

Figure 2.2A–D illustrates the behavior of the ternary complex model when there is an excess of G protein ($G_T/R_T = 4$) and a lack of constitutive activity ($Kc << 1$). Theoretical curves are shown for a highly efficacious agonist (Figure 2.2A), a less efficacious agonist (Figure 2.2B), an antagonist (Figure 2.2C), and an intrinsic antagonist with selectivity for the ground state (Figure 2.2D). In each case, receptor occupancy is nearly consistent with a simple one-site model as shown by the similarity in the observed dissociation constants of the *DRG* and *DR* complexes (see Table 2.2). For both the highly efficacious agonist and the less efficacious agonist, a substantial amount of the bound agonist is in the form of the ternary complex (*DRG*), with the maximal amount formed being correlated with α. The value of α increases as the ratio of affinity constants of the ground and active states of the receptor (a) increases. The parameter αK_2 is expected to be proportional to intrinsic efficacy as described below. For the two types of antagonist, the bound form of the receptor is attributed almost entirely to the *DR* complex.

Conformational analysis of the curves shown in Figure 2.2 is summarized in Table 2.2. It can be seen that, for the agonists, the dominant conformation of the ternary complex is $DR_s{*}G$, whereas the dominant conformation of the drug receptor complex is DR_s. For the antagonists, the dominant conformation of bound drug is DR_s. The observed dissociation constant (K_{obs}) of all of the drug–receptor conformations is the same.

It is interesting to compare the observed binding curve of an agonist with its microscopic affinity constants for the receptor complexes and conformations. Such a comparison is shown in Figure 2.3 for the highly efficacious agonist described in Figure 2.2A. It can be seen that the occupancy curve for the agonist is determined by its observed dissociation constant (K_{obs}), and that this observed affinity is much less than its affinity for the ternary complex (αK_1) or the active state of the receptor (aK_a). This condition arises because the receptor is in a conformation that the agonist does not prefer (i.e., the ground state). Consequently, part of the intrinsic binding energy of the agonist is used to induce a conformational change in the receptor.[12] In contrast, antagonists lack selectivity for the active state; consequently, the observed affinity is approximately equal to its affinity for *DR* and R_s (i.e., K_1 and K_a).

Figure 2.4A–D shows the behavior of the ternary complex model when the receptor exhibits constitutive activity. This activity arises when the value of K_c is increased, that is, when more of the receptor is in the active conformation in the absence of agonist. Increasing the constitutive activity of the receptor increases the observed affinities of the agonists (Figures 2.4A and B) while reducing the observed affinity of the intrinsic antagonist (Figure 2.4D). Receptor activation is proportional to the amount of receptor in the active conformation. Conformational analysis shows that the active conformation of the receptor exists almost entirely in the guanine nucleotide coupled complex (i.e., *RG* + *DRG*). In contrast to Figure 2.2, where the *DRG* complex accounted for essentially all of the G protein-coupled form of the receptor, a substantial amount of the G protein is coupled to the unoccupied receptor (i.e., *RG*) under the conditions of Figure 2.4. Thus, Figure 2.4 shows the total amount of receptor in the G protein-coupled complex (*RG* + *DRG*) as well as the receptor

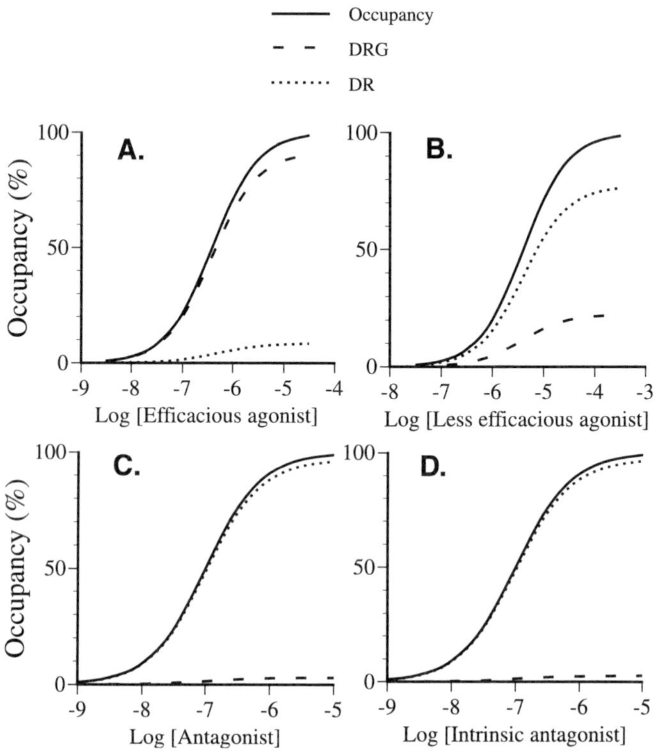

FIGURE 2.2 Conformational analysis of the ternary complex model for the binding of drugs to a receptor with excess G protein ($G_T/R_T = 4$) *and little constitutive activity.* The solid lines show total receptor occupancy, whereas the two types of dashed lines show the proportion of drug bound in the form of ternary complex (*DRG*) and agonist–receptor complex (*DR*). The model is fully defined at the three hierarchical levels corresponding to the observed binding constants (see Table 2.2), the microscopic constants of the receptor complexes, and the microscopic constants of the receptor states. The microscopic state constants for this receptor-G protein system are: $K_b = 7 \times 10^{-3}$, $K_c = 10^{-6}$, and $b = 10^5$. These microscopic state constants determine the microscopic constant for the receptor-G protein complex: $K_2 = 7.7 \times 10^{-3}$. The ligand-dependent microscopic constants are given in connection with each panel of the figure as follows. A. The binding of an efficacious agonist. The microscopic state constants are: $K_a = 2 \times 10^5$ M^{-1} and $a = 5 \times 10^3$. The microscopic constants for the ligand–receptor complexes are: $K_1 = 2.01 \times 10^5$ M^{-1} and $\alpha = 4.53 \times 10^2$. B. The binding of a less efficacious agonist. The microscopic state constants are: $K_a = 2 \times 10^5$ M^{-1} and $a = 10^2$. The microscopic constants for the ligand–receptor complexes are: $K_1 = 2 \times 10^5$ M^{-1} and $\alpha = 10$. C. The binding of an antagonist. The microscopic state constants are: $K_a = 10^7$ M^{-1} and $a = 1.0$. The microscopic constants for the ligand–receptor complexes are: $K_1 = 10^7$ M^{-1} and $\alpha = 1.0$. D. The binding of an intrinsic antagonist. The microscopic state constants are: $K_a = 10^7$ M^{-1} and $a = 10^{-4}$. The microscopic constants for the ligand–receptor complexes are: $K_1 = 10^7$ M^{-1} and $\alpha = 9.09 \times 10^{-1}$.

TABLE 2.2
Conformational Analysis of the Ternary Complex Model for the Binding of Drugs to a Receptor with Excess G Protein and Little Constitutive Activity. The binding parameters were estimated from the theoretical data shown in Figure 2.3.

	Agonist, Efficacious	Agonist, Less Efficacious	Antagonist	Antagonist, Intrinsic
		DRG Complex		
pK_{DRG}[a]	6.42	5.42	7.00	7.00
$DR_s{*}G_{max}$[b]	91.3%	20.5%	0.27%	<.01%
DR_sG_{max}[c]	0.18%	2.05%	2.70%	2.71%
		DR Complex		
pK_{DR}[d]	6.31	5.34	7.00	7.00
$DR_s{*}_{max}$[e]	0.04%	0.01%	<.01%	<.01%
$DR_{s\ max}$[f]	8.46%	77.5%	97.0%	97.3%

[a] Negative logarithm of the concentration of drug required for half-maximal occupancy of the DRG complex.
[b] Maximal amount of $DR_s{*}G$ complex formed at 100% receptor occupancy.
[c] Maximal amount of DR_sG complex formed at 100% receptor occupancy.
[d] Negative logarithm of the concentration of drug required for half-maximal occupancy of the *DR* complex.
[e] Maximal amount of $DR_s{*}$ complex formed at 100% receptor occupancy.
[f] Maximal amount of DR_s complex formed at 100% receptor occupancy.

occupancy. It can be seen that a substantial amount of receptor is precoupled to G protein before the agonist is present, and that the maximal amount of receptor-G protein complex formed is correlated with the ratio of the microscopic affinity constants of the ground and active conformations of the receptor (a). In contrast, the antagonist that has equal affinity for the ground and active states has no influence on the amount of receptor in the receptor-G protein complex, whereas the intrinsic antagonist reduces the amount of receptor-G protein complex.

Conformational analysis of the data illustrated in Figure 2.4 is shown in Table 2.3. Once again, for the agonists, most of the *DRG* complex is attributed to the $DR_s{*}G$ conformation, whereas the *DR* complex is mainly in the form of DR_s. For the antagonist lacking selectivity for the ground and active conformations, the distribution of bound drug in the form of the ground and active conformations (DR_s, $DR_s{*}G$) is essentially the same as that observed in the basal state. For the intrinsic antagonist, nearly the entire fraction of bound drug is in the DR_s conformation. Most of the precoupled receptor-G protein complex (*RG*) is in the $R_s{*}G$ conformation.

It appears to be a popular misconception that the simple ternary complex model cannot account for constitutively active receptors.[13] However, this is clearly not the case. Constitutive activity occurs when there is a sufficient amount of receptor in

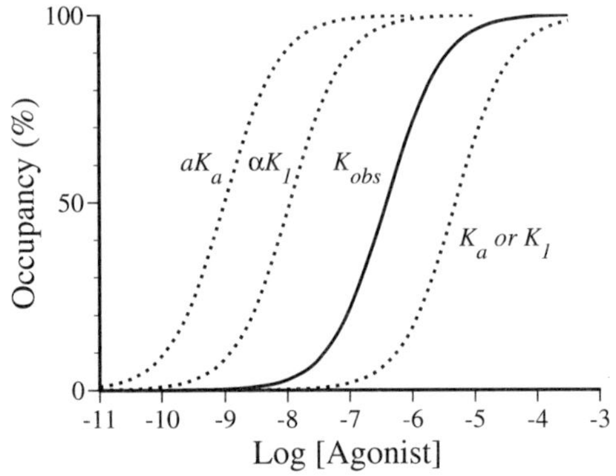

FIGURE 2.3 A comparison of the three different types of dissociation constants in the ternary complex model for the efficacious agonist shown in Figure 2.2A. The solid curve is a true measure of receptor occupancy and is the same as that shown in Figure 2.2A. This curve is nearly consistent with a simple one-site model because there is an excess of G protein to receptor (i.e., $G_T/R_T = 4.0$). The concentration of agonist yielding 50% receptor occupancy equivalent to K_{obs}. Under these conditions, the K_{obs} of the *DRG* and *DR* complexes are approximately the same (see Table 2.2). The dotted lines show the imaginary occupancy curves if the receptor was locked into the *DRG* complex (αK_1), the *DR* complex (K_1), the active conformation (aK_a), or the ground conformation (K_a).

the active conformation so that the receptor elicits a response in the absence of agonist. At the level of conformational analysis, this situation occurs when the value of K_c increases to some critical level. However, Equation (6) shows that as the value of K_c increases, so does the value of K_2. Thus, at the level of the simple ternary complex model, constitutive activity is characterized by an increase in the value of K_2, which causes coupling of the receptor to the G protein in the absence of agonist.

2.2.3 Behavior of the Ternary Complex Model with Limiting G Protein

G proteins increase the observed affinity of the agonist for the receptor. Consequently, when the G protein concentration is limiting (e.g., $G_T < R_T$), there may not be enough of it to enhance the affinity of agonists for all of the receptors. Consequently, some of the receptors behave as if they have lower affinity for agonists. Under these conditions, the binding curves of agonists exhibit a low slope, similar to that expected for a two-site binding curve.[2]

Figure 2.5A–D shows the theoretical binding curves of agonists and antagonists when there is a limiting amount of G protein (i.e., $G_T/R_T = 0.8$) and the receptor lacks constitutive activity. For the highly efficacious agonist (Figure 2.5A), binding is generally consistent with a two-site model, with the high and low affinity components composed of the *DRG* and *DR* complexes, respectively. Similar behavior is observed for the partial agonist (Figure 2.5B); however, there is less of an observed

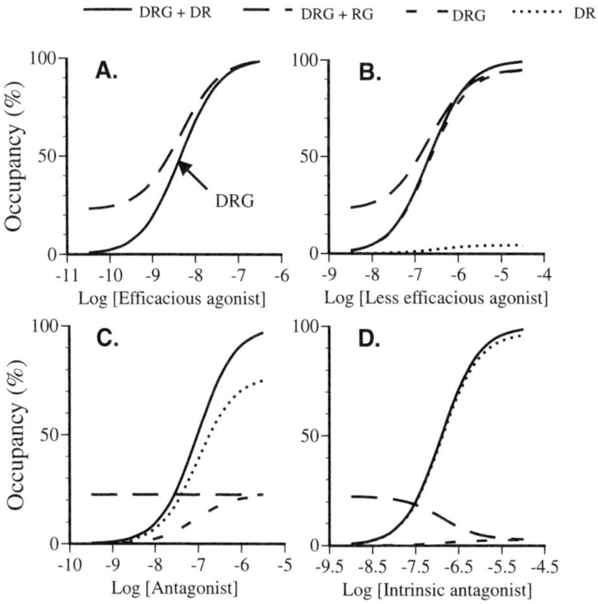

FIGURE 2.4 Conformational analysis of the ternary complex model for the binding of drugs to a constitutively active receptor with excess G protein (G_T/R_T = 4). The solid lines show total receptor occupancy, whereas the three types of dashed lines show the proportion of drug bound in the form of ternary complex (*DRG*), agonist-receptor complex (*DR*), and the summation of the G protein-coupled forms of the receptor (*DRG* + *RG*). The model is fully defined at the three hierarchical levels corresponding to the observed binding constants (see Table 2.2), the microscopic constants of the receptor complexes, and the microscopic constants of the receptor states. The microscopic state constants for this receptor-G protein system are: $K_b = 7 \times 10^{-3}$, $K_c = 10^{-4}$, and $b = 10^5$. These microscopic state constants determine the microscopic constant for the receptor-G protein complex: $K_2 = 7.7 \times 10^{-2}$. The ligand-dependent microscopic constants for the receptor conformations are the same as those given in the caption for Figure 2.1A, whereas the ligand-dependent constants for the receptor complexes are given in connection with each panel of the figure as follows. A. The binding of an efficacious agonist. The microscopic constants for the ligand–receptor complexes are: $K_1 = 3 \times 10^5$ M^{-1} and $\alpha = 3.03 \times 10^3$. The *DRG* complex accounts for practically all of the bound drug; consequently, the occupancy curves for *DRG* and the summation of *DRG* and *DR* (total receptor occupancy) are superimposible. Moreover, very little *DR* complex is formed. B. The binding of a less efficacious agonist. The microscopic constants for the ligand–receptor complexes are: $K_1 = 2.02 \times 10^5$ M^{-1} and $\alpha = 90.1$. C. The binding of an antagonist. The microscopic constants for the ligand–receptor complexes are: $K_1 = 10^7$ M^{-1} and $\alpha = 1.0$. D. The binding of an intrinsic antagonist. The microscopic constants for the ligand-receptor complexes are: $K_1 = 10^7$ M^{-1} and $\alpha = 9.09 \times 10^{-1}$.

TABLE 2.3
Conformational Analysis of the Ternary Complex Model for the Binding of Drugs to a Constitutively Active Receptor with Excess G Protein. The binding parameters were estimated from the theoretical data shown in Figure 2.4.

	Agonist, Efficacious	Agonist, Less Efficacious	Antagonist	Antagonist, Intrinsic
		DRG Complex		
pK_{DRG}[a]	8.34	6.67	7.06	6.88
$DR_s{*}G_{max}$[b]	99.9%	95.4%	20.5%	<.01%
DR_sG_{max}[c]	<.01%	0.10%	2.05%	2.71%
		DR Complex		
pK_{DR}[d]	8.17	6.50	7.00	6.90
$DR_s{*}_{max}$[e]	0.05%	0.04%	0.01%	<.01%
$DR_{s\ max}$[f]	0.10%	4.47%	77.5%	97.3%
		RG Complex		
pIC_{50}[g]	8.34	6.66	7.00	6.88
$R_s{*}G_{initial}$[h]	20.5%	20.5%	20.5%	20.5%
$R_sG_{initial}$[i]	2.05%	2.05%	2.05%	2.05%

[a] Negative logarithm of the concentration of drug required for half-maximal occupancy of the *DRG* complex.
[b] Maximal amount of $DR_s{*}G$ complex formed at 100% receptor occupancy.
[c] Maximal amount of DR_sG complex formed at 100% receptor occupancy.
[d] Negative logarithm of the concentration of drug required for half-maximal occupancy of the *DR* complex.
[e] Maximal amount of $DR_s{*}$ complex formed at 100% receptor occupancy.
[f] Maximal amount of DR_s complex formed at 100% receptor occupancy.
[g] Negative logarithm of the concentration of drug required for half-maximal inhibition of the *RG* complex.
[h] Initial amount of $R_s{*}G$ complex formed in the absence of drug.
[i] Initial amount of R_sG complex formed in the absence of drug.

affinity difference between the *DRG* and *DR* components, and the contribution of the *DRG* component is less. In contrast, the antagonist (Figure 2.5C) and the intrinsic antagonist (Figure 2.5D) both exhibit simple, one-site binding curves composed mainly of the *DR* complex.

The binding parameters of Figure 2.5 are summarized in Table 2.4. The negative logarithm of the concentration of agonist required to generate half-maximal formation of the *DRG* and *DR* complexes (pK_{DRG} and pK_{DR}, respectively) are listed in Table 2.4. It can be seen that, for agonists, the observed affinity of the ternary complex (pK_{DRG}) is actually much less than the microscopic affinity constant of the ternary complex (αK_1; see legend to Figure 2.5). In contrast, the observed affinity of the *DR* complex (pK_{DR}) is actually a little greater than the microscopic affinity constant of the *DR* complex (K_1; see legend to Figure 2.5).

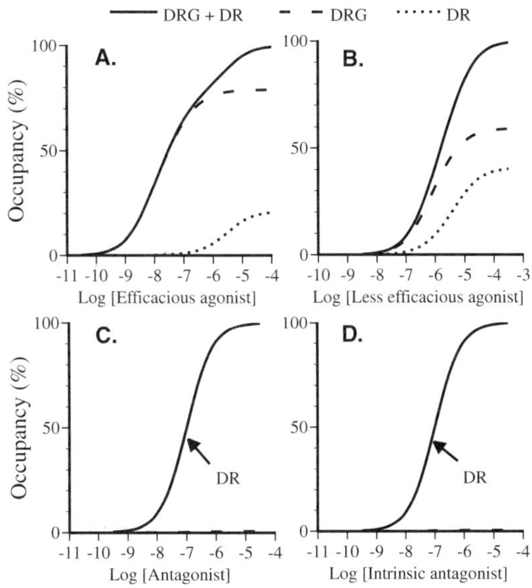

FIGURE 2.5 Conformational analysis of the ternary complex model for the binding of drugs to a receptor with limiting G protein ($G_T/R_T = 0.8$) and little constitutive activity. The solid lines show total receptor occupancy, whereas the two types of dashed lines show the proportion of drug bound in the form of ternary complex (*DRG*) and agonist–receptor complex (*DR*). The model is fully defined at the three hierarchical levels corresponding to the observed binding constants (see Table 3.4), the microscopic constants of the receptor complexes, and the microscopic constants of the receptor states. The microscopic state constants for this receptor-G protein system are: $K_b = 7 \times 10^{-3}$; $K_c = 10^{-6}$; and $b = 10^5$. These microscopic state constants determine the microscopic constant for the receptor-G protein complex: $K_2 = 7.7 \times 10^{-3}$. The ligand-dependent microscopic constants are given in connection with each panel of the figure as follows. A. The binding of an efficacious agonist. The microscopic state constants are: $K_a = 1.6 \times 10^5$ M^{-1} and $a = 10^6$. The microscopic constants for the ligand–receptor complexes are: $K_1 = 3.2 \times 10^5$ M^{-1} and $\alpha = 4.55 \times 10^4$. B. The binding of a less efficacious agonist. The microscopic state constants are: $K_a = 1.6 \times 10^5$ M^{-1} and $a = 10^4$. The microscopic constants for the ligand–receptor complexes are: $K_1 = 1.62 \times 10^5$ M^{-1} and $\alpha = 9.01 \times 10^2$. C. The binding of an antagonist. The microscopic state constants are: $K_a = 10^7$ M^{-1} and $a = 1.0$. The microscopic constants for the ligand–receptor complexes are: $K_1 = 10^7$ M^{-1} and $\alpha = 1.0$. D. The binding of an intrinsic antagonist. The microscopic state constants are: $K_a = 10^7$ M^{-1} and $a = 10^{-4}$. The microscopic constants for the ligand–receptor complexes are: $K_1 = 10^7$ M^{-1} and $\alpha = 9.09 \times 10^{-1}$. In panels C and D, the *DR* complex accounts for nearly all of the bound drug; consequently, the receptor occupancy curves for *DR* and the summation of *DRG* and *DR* (total receptor occupancy) are superimposable.

Conformational analysis is also summarized in Table 2.4. As observed for agonists in Figures 2.2 and 2.4, most of the *DRG* complex is in the $DR_s{}^*G$ state, whereas the DR_s state accounts for most of *DR* complex. For the antagonists, the DR_s state accounts for nearly the entire fraction of *DR* complex. It can also be seen that the observed affinities of the DR_sG and $DR_s{}^*G$ conformations are the same (i.e., pK_{DRG}) as well as those of the DR_s and $DR_s{}^*$ conformations (i.e., pK_{DR}).

TABLE 2.4
Conformational Analysis of the Ternary Complex Model for the Binding of Drugs to a Receptor with Limiting G Protein and Little Constitutive Activity. The binding parameters were estimated from the theoretical data shown in Figure 2.5.

	Agonist, Efficacious	Agonist, Less Efficacious	Antagonist	Antagonist, Intrinsic
		DRG Complex		
pK_{DRG} [a]	7.82	6.06	7.00	7.00
$DR_s{*}G_{max}$ [b]	78.9%	59.1%	0.06%	<.01%
DR_sG_{max} [c]	<.01%	.06%	0.55%	.55%
		DR Complex		
pK_{DR} [d]	5.55	5.42	7.00	7.00
$DR_s{*}_{max}$ [e]	10.5%	0.41%	<.01%	<.01%
$DR_{s\ max}$ [f]	10.5%	40.5%	99.4%	99.4%

[a] Negative logarithm of the concentration of drug required for half-maximal occupancy of the *DRG* complex.
[b] Maximal amount of $DR_s{*}G$ complex formed at 100% receptor occupancy.
[c] Maximal amount of DR_sG complex formed at 100% receptor occupancy.
[d] Negative logarithm of the concentration of drug required for half-maximal occupancy of the *DR* complex.
[e] Maximal amount of $DR_s{*}$ complex formed at 100% receptor occupancy.
[f] Maximal amount of DR_s complex formed at 100% receptor occupancy.

Figure 2.6A–D shows the behavior of the ternary complex model when the G protein is limiting and there is constitutive activity. The conditions of Figure 2.6 are similar to those of Figure 2.5 except that the value of K_c was increased to induce constitutive activity. As observed in Figure 2.4, constitutive activity increases the observed affinity of agonists while reducing the observed affinity of the intrinsic antagonist. The binding parameters and conformational analysis of Figure 2.6 are summarized in Table 2.5. The behavior of agonists can be rationalized as the result of a model that combines constitutive activity (Table 2.3) with a limiting supply of G protein (Table 2.4). An interesting result occurs with the intrinsic antagonist. The binding curve resembles a two-site model. However, in contrast to that observed with agonists, the high affinity component is composed of the *DR* complex, whereas the low affinity component consists of the *DRG* complex. Conformational analysis shows that the *DR* complex consists mainly of the DR_s conformation whereas the *DRG* complex consists mainly of the $DR_s{*}G$ conformation.

2.2.4 Relevance of the Ternary Complex Model to Ligand Binding Data

G protein-linked receptors often display GTP-sensitive, two-site binding curves for agonists and GTP-insensitive, one-site curves for antagonists. This behavior is gen-

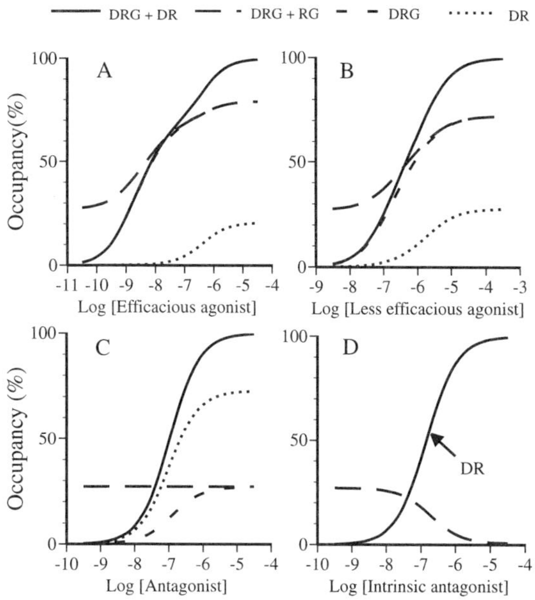

FIGURE 2.6 Conformational analysis of the ternary complex model for the binding of drugs to a constitutively active receptor with limiting G protein ($G_T/R_T = 0.8$). The solid lines show total receptor occupancy, whereas the three types of dashed lines show the proportion of drug bound in the form of ternary complex (*DRG*), agonist-receptor complex (*DR*), and the summation of the G protein-coupled forms of the receptor (*DRG* + RG). The model is fully defined at the three hierarchical levels corresponding to the observed binding constants (see Table 2.5), the microscopic constants of the receptor complexes, and the microscopic constants of the receptor states. The microscopic state constants for this receptor-G protein system are: $K_b = 7 \times 10^{-3}$, $K_c = 10^{-3}$, and $b = 10^5$. These microscopic state constants determine the microscopic constant for the receptor-G protein complex: $K_2 = 0.71$. The ligand-dependent microscopic constants are given in connection with each panel of the figure as follows. A. The binding of an efficacious agonist. The microscopic state constants are: $K_a = 4 \times 10^5$ M^{-1} and $a = 5 \times 10^3$. The microscopic constants for the ligand–receptor complexes are: $K_1 = 2.4 \times 10^6$ M^{-1} and $\alpha = 8.26 \times 10^2$. B. The binding of a less efficacious agonist. The microscopic state constants are: $K_a = 4 \times 10^5$ M^{-1} and $a = 50$. The microscopic constants for the ligand–receptor complexes are: $K_1 = 4.2 \times 10^5$ M^{-1} and $\alpha = 47.2$. C. The binding of an antagonist. The microscopic state constants are: $K_a = 10^7$ M^{-1} and $a = 1.0$. The microscopic constants for the ligand–receptor complexes are: $K_1 = 10^7$ M^{-1} and $\alpha = 1.0$. D. The binding of an intrinsic antagonist. The microscopic state constants are: $K_a = 10^7$ M^{-1} and $a = 10^{-4}$. The microscopic constants for the ligand–receptor complexes are: $K_1 = 10^7$ M^{-1} and $\alpha = 10^{-2}$. In panel D, the *DR* complex accounts for nearly all of the bound drug; consequently, the occupancy curves for the *DR* complex and the summation of *DR* and *DRG* (total receptor occupancy) are superimposable.

TABLE 2.5
Conformational Analysis of the Ternary Complex Model for the Binding of Drugs to a Constitutively Active Receptor with Limiting G Protein. The binding parameters were estimated from the theoretical data shown in Figure 2.6.

	Agonist, Efficacious	Agonist, Less Efficacious	Antagonist	Antagonist, Intrinsic
		DRG Complex		
pK_{DRG}[a]	8.41	6.56	6.91	6.67
$DR_s{*}G_{max}$[b]	79.3%	72.2%	26.9%	0.01%
DR_sG_{max}[c]	<.01%	0.01%	0.27%	0.06%
		DR Complex		
pK_{DR}[d]	6.41	5.79	7.00	6.84
$DR_s{*}_{max}$[e]	17.2%	1.32%	0.07%	<.01%
$DR_{s\ max}$[f]	3.44%	26.5%	72.8%	99.4%
		RG Complex		
pIC_{50}[g]	8.93	6.95	7.00	6.71
$R_s{*}G_{initial}$[h]	26.9%	26.9%	26.9%	26.9%
$R_sG_{initial}$[i]	0.27%	0.27%	0.27%	0.27%

[a] Negative logarithm of the concentration of drug required for half-maximal occupancy of the *DRG* complex.
[b] Maximal amount of $DR_s{*}G$ complex formed at 100% receptor occupancy.
[c] Maximal amount of DR_sG complex formed at 100% receptor occupancy.
[d] Negative logarithm of the concentration of drug required for half-maximal occupancy of the *DR* complex.
[e] Maximal amount of $DR_s{*}$ complex formed at 100% receptor occupancy.
[f] Maximal amount of DR_s complex formed at 100% receptor occupancy.
[g] Negative logarithm of the concentration of drug required for half-maximal inhibition of the *RG* complex.
[h] Initial amount of $R_s{*}G$ complex formed in the absence of drug.
[i] Initial amount of R_sG complex formed in the absence of drug.

erally consistent with that predicted by the ternary complex model with limiting G protein (see Figure 2.5). Nevertheless, there is reason to question how well this condition of the model accurately describes data of this sort. The main problem is that G proteins, particularly G_i and G_o, often outnumber their receptors by as much as several hundredfold.[14,15] Why then does agonist binding exhibit heterogeneity if the ternary complex model predicts one-site binding behavior under these conditions (see Figures 2.2 and 2.4)? There is no clear answer, but several factors could contribute to a local shortage of G protein in the vicinity of the receptor.

If there were at least two pools of receptors — one corresponding to a membrane fraction enriched in G protein and the other devoid of G protein — then behavior consistent with a model of two independent binding sites would result.[16] These two sites would exhibit differential affinity for agonists and the same affinity for antagonists. If this model were correct, then the proportion of high affinity sites should be constant for all agonists. However, in many instances the proportion of high affinity sites is correlated with the relative efficacy of agonists. Consequently, the model of independent pools of receptors cannot entirely explain the data, although it does seem likely that tissue homogenates are heterogeneous and contain membrane particles that vary in their composition of G protein and receptor.

It is possible that constraints imposed by the two-dimensional nature of the membrane limit the interaction of G protein and receptor, giving rise to shallow binding curves for agonists.[17] Also, there may be a requirement for a scaffold-like protein to facilitate receptor-G protein interactions. The mathematics of the ternary complex model does not directly take into account these possibilities. The model uses equilibrium constants that are appropriate for ligands and proteins that diffuse along three dimensions to interact with each other. Consequently, the only way for the model to account for a limitation in the supply of G protein is through a reduction in the ratio of G_T/R_T. Therefore, the ratio G_T/R_T in the model is probably not a true ratio, but rather a parameter defined by microscopic constants including the true concentrations of receptor, G protein, and other factors (e.g., membrane constraints, scaffolding proteins, heterogeneous distribution of receptors, and G proteins).

Although the agonist binding curves shown in Figures 2.5 and 2.6 have two components, they actually deviate somewhat from that predicted by a model of two independent binding sites. The deviation can be attributed to the high affinity component of the ternary complex model, which has a Hill slope of less than one. This behavior may explain why several investigators have reported that it is necessary to consider three types of sites — superhigh, high, and low affinity (SH, H, and L) — to account for the binding of agonists to cardiac M_2 muscarinic receptors adequately.[18,19] Since the high affinity component of the ternary complex model has a Hill slope of less than one, it can be represented arbitrarily as the summation of two sites (SH and H). In contrast, the *DR* component of the ternary complex model has a Hill slope of one and is consistent with a single low affinity site. Collectively, therefore, the two components of the ternary complex model yield a binding curve that requires three sites to describe it adequately when modeled from the perspective of independent binding sites.[17] These results underscore the importance of a mechanistic approach to modeling.

2.2.5 Analysis of Ligand Binding Data with the Ternary Complex Model

Although innumerable binding experiments on G protein-linked receptors have been described, few of these reports describe the fundamental parameters of the ternary complex model. Usually, binding data are analyzed in terms of a model of independent binding sites, and intuitive conclusions about the ternary complex model are made. When a two-site model is used to analyze data similar to that shown in Figure 2.5, the estimates of the high and low affinity constants correspond to the observed affinity of the agonist for the ternary complex and the drug receptor complex, respectively. However, these estimates are not equivalent to the microscopic affinity constant of the agonist for the ternary complex (αK_1) and drug receptor complex (K_1).[20] Nevertheless, the observed proportion of high affinity sites is a reasonable approximation of the maximal amount of ternary complex formed at high concentrations of the agonist.

In binding assays, direct measurements of ligand–receptor complexes are made. It is usually not possible to discriminate between different conformations of a receptor complex because they have similar observed affinities (see Figures 2.5 and 2.6). For example, the high affinity receptor complex for the agonists shown in Figure 2.5 corresponds to the ternary complex (*DRG*). The two conformations of this complex (DR_sG and DR_s*G) have the same observed affinity (pK_{DRG}). Consequently, it is not possible to discriminate between these two conformations; it is only possible to discriminate between the *DRG* and *DR* complexes. The latter complex also has two conformations (DR_s and DR_s*); however, both of these also have the same observed affinity. Therefore, the appropriate level of analysis for investigating ligand binding data is at the level of the simple ternary complex model (second hierarchical level of analysis). It is not possible to estimate the parameters in the expanded form of the model from a single binding curve. Moreover, the simple ternary complex model can account for all of the behaviors of the expanded model, including constitutive activity (see Figures 2.4 and 2.6).

In order to estimate all of the parameters of the ternary complex model, it is necessary to measure a series of agonist binding curves at different ratios of G protein to receptor. In principle, this information should be sufficient to estimate the microscopic affinity constants of the agonist for the *DRG* and *DR* complexes (i.e., αK_1 and K_1) as well as the microscopic affinity constants of the G protein for the *DRG* and *RG* complexes (i.e., αK_2 and K_2). The ratio of G protein to receptor could be manipulated artificially by using a system that has a relatively low ratio of G_T/R_T to begin with, and then inactivating some of the receptors with an irreversible antagonist. Receptor inactivation has the effect of increasing the ratio of G protein to receptor, which should increase the overall affinity of the agonist and alter the slope of its binding curve. In other words, receptor inactivation should convert an agonist binding curve from the G protein limiting situation shown in Figure 2.5 to the G protein excess condition shown in Figure 2.2. However, when experiments of this type were carried out on myocardial M_2 receptors, receptor inactivation was shown to have no effect on the overall affinity of the agonist or on the slope of its binding curve (unpublished observations). On the surface, these results seem to invalidate the simple ternary complex model. However, as described above, perhaps the constraints of the two-dimensional membrane as well as the loading of receptor and G protein onto potential scaffolding proteins limits access of G proteins to the receptor despite a relative excess of the G protein. Under conditions of limited access, an inactive receptor behaves in some ways like a dominant negative mutant, hindering the interaction of the residual active receptors with G proteins. An alternative approach to altering the ratio G_T/R_T for G_i- and G_o-linked receptors is to treat the receptor preparation with pertussis toxin. Pertussis toxin prevents receptor interactions with the G_i and G_o family of G proteins.[21-23] In agreement with the ternary complex model, pertussis toxin treatment reduces the overall affinity of agonists at a variety of G_i- and G_o-linked receptors and tends to convert the two-site agonist binding curve into one of more uniform low affinity.[21,22,24]

A method for estimating some of the parameters of the ternary complex model has been previously described for M_2 muscarinic receptors in the rabbit myocardium.[20] In this prior study, the mammalian myocardium was used because it expresses a nearly uniform population of M_2 receptors.[25,26] The method of analysis has been described in detail previously, and a summary of the technique is given below.

Since agonists typically bind to G protein-linked receptors over a concentration range that can exceed several log units, it is necessary to measure agonist binding indirectly by competitive inhibition of the binding of a fixed concentration of a labeled antagonist. To simplify the analysis, an antagonist that exhibits uniform binding properties should be used. For M_2 muscarinic receptors in the heart, [^{3}H]N-methylscopolamine ([^{3}H]NMS) satisfies these criteria provided that the assays are run in the presence of a physiological buffer system. Under these conditions, [^{3}H]NMS binding is consistent with a simple one-site model and is unaffected by guanine nucleotides. However, in low ionic strength buffer, [^{3}H]NMS binding exhibits two-site behavior, and its overall affinity is reduced.[27] Under these latter conditions, guanine nucleotides increase the observed affinity of [^{3}H]NMS and convert its binding curve into one of uniform high affinity. These results can be explained by the precoupling of M_2 receptors to G proteins in low ionic strength buffer in the absence of agonist. Under these conditions, the M_2 receptor exhibits constitutive activity with regard to [^{35}S]GTPγS binding, and NMS and other muscarinic antagonists behave as intrinsic antagonists and inhibit this constitutive activity.[28]

Preliminary analysis of the competitive binding curves of muscarinic agonists in the heart showed that an infinite number of parameter estimates (i.e., G_T/R_T, α, K_1 and K_2) could provide a good fit to the data. Nevertheless, it was possible to estimate G_T/R_T, K_1 and the product αK_2, but not individual estimates of α and K_2.[20] To obtain these estimates, the ternary complex model was fitted to the data by nonlinear regression analysis, with the value of K_2 fixed to various values between the range of 10 and 10^{-8}. Regression analysis showed that as the value of K_2 decreased to less than 10^{-2}, the residual sum of the squared deviations (RSS) between the best fitting curve and the predicted binding values approached a minimum. Moreover, this minimum remained constant as the value of K_2 was reduced to 10^{-8}. Similarly, when the value of K_2 was less than 10^{-2}, the estimates of G_T/R_T and K_1 also reached a plateau at constant values. In contrast, when the value of K_2 was equal to or less than 10^{-2}, the value of α increased proportionately with the reciprocal of K_2 so that the product (αK_2) of the two estimates remained constant. Therefore, it was possible to obtain the best estimates of G_T/R_T, K_1 and αK_2 by setting K_2 to an arbitrarily low value (10^{-4}) and determining the parameter estimates that minimized RSS. Simultaneous regression analysis of the competition data of nine agonists showed that there was no significant increase in residual error when the value of G_T/R_T was shared among the various sets of data. The best estimate of G_T/R_T was 0.81, and the estimates of K_1 and αK_2 from this study are listed in Table 2.6. Once again, though, the estimate of G_T/R_T is unreliable. This parameter is probably a macroscopic constant that reflects the G protein concentration as well as other factors that limit the access of the G

TABLE 2.6
The Binding Parameters of Agonists for M_2 Muscarinic Receptors in the Rabbit Myocardium. The binding data were analyzed according to the simple ternary complex model. The data are from Ehlert[20] with permission.

Agonist	Log K_1	αK_2
Oxotremorine-M	5.5	381
Carbachol	4.1	187
cis-Dioxolane	4.8	157
Oxotremorine	6.2	95
(+)-Aceclidine	4.7	66
(–)-Aceclidine	4.6	46
N-Methylaceclidine	4.1	22
BM5	6.3	17
BOK1	6.6	8.6

protein to the receptor (see above). As described in a later section, there was a strong correlation between the estimate of αK_2 and the relative efficacy of agonists.

2.3 BEHAVIOR OF THE TERNARY COMPLEX MODEL IN THE PRESENCE OF GUANINE NUCLEOTIDES

Since the cytosolic concentration of GTP is in the near millimolar range,[3-5] no description of the ternary complex model is complete without an analysis of how the model behaves in the presence of guanine nucleotide. This issue is important because classic receptor theory has relevance to functioning receptors, and G protein-linked receptors normally function in the presence of concentrations of GTP that saturate the G protein. Thus, in order to evaluate classic receptor theory in terms of the ternary complex model, we need to consider the model with guanine nucleotide present. We will begin by first examining the allosteric transitions that are associated with G protein subunit association and dissociation.

2.3.1 Influence of Guanine Nucleotide on G Protein Quaternary Structure

The G_α subunit of G proteins can be likened to a receptor with two allosterically linked binding sites — one for guanine nucleotide (X) and the other for $G_{\beta\gamma}$.[29,30] The binding of guanine nucleotide influences the binding of $G_{\beta\gamma}$, and conversely, the binding of $G_{\beta\gamma}$ influences the binding of guanine nucleotide. These interactions are described in the following scheme:

$$
\begin{array}{ccc}
 & K_4 & \\
X + G_\alpha + G_{\beta\gamma} & \longleftrightarrow & XG_\alpha + G_{\beta\gamma} \\
K_5 \updownarrow & & \updownarrow \pi K_5 \\
X + G_\alpha G_{\beta\gamma} & \longleftrightarrow & XG_\alpha G_{\beta\gamma} \\
 & \pi K_4 &
\end{array}
\qquad (13)
$$

in which K_4 denotes the microscopic affinity constant for the binding of guanine nucleotide (X) to G_α, K_5 denotes the microscopic affinity constant for the binding of $G_{\beta\gamma}$ to G_α, and π denotes the cooperativity. These interactions are analogous to those of the simple ternary complex model shown in scheme 1, and thus, the system behaves in an analogous fashion. When GTP (X_T) binds to G_α, it causes a dissociation of the $G_{\beta\gamma}$ subunit. Consequently, there is negative heterotropic cooperativity between the binding of GTP and the $G_{\beta\gamma}$ subunit ($0 < \pi < 1$). In contrast, when GDP (X_D) binds to G_α, it promotes the association of G_α and $G_{\beta\gamma}$. Thus, there is positive heterotropic cooperativity between the binding of GDP and $G_{\beta\gamma}$ ($\pi > 1$). These allosteric changes in G protein subunits have been monitored through measurement of their intrinsic fluorescence.[31,32]

2.3.2 Ternary Complex Model with Guanine Nucleotide

To describe a functioning receptor with the ternary complex model, we need to add guanine nucleotide to the model. Since GTP is maintained at high concentrations inside the cell, a useful starting point is to imagine that all of the G proteins are bound with GTP. The GTP complex would spontaneously dissociate into its GTP-G_α and $G_{\beta\gamma}$ subunits, causing a huge burst of cellular stimulation as the free subunits activated their respective effector enzymes and ion channels. However, this artificial starting condition would rapidly decay because of the intrinsic GTPase activity of the G_α subunit. Soon, most of the GTP-G_α would be converted to GDP-G_α, causing the recombining of the GDP-G_α and $G_{\beta\gamma}$ subunits to form the GDP-bound form of the G protein (*G-GDP*). Very little of the large amount of GTP in the cytosol would have the opportunity to displace GDP from the G protein because the rate of dissociation of GDP is so slow relative to the GTPase activity. Consequently, under basal conditions, most of the G proteins are bound with GDP in an inactive state. This condition represents a steady-state, which is driven far from equilibrium by the GTPase activity of the G protein (see Figure 2.7). The role of the agonist–receptor complex is to increase the dissociation rate of GDP from the G protein. This effect is most likely manifested as an agonist-induced reduction in the affinity of GDP for the G protein. This enables GTP to displace GDP rapidly, resulting in activation and a much larger proportion of the G proteins occupied by GTP. Thus, the agonist receptor complex pushes the system away from the basal steady-state and closer to an equilibrium state, where the high concentration of GTP is the determining factor in competing GDP off the G protein.

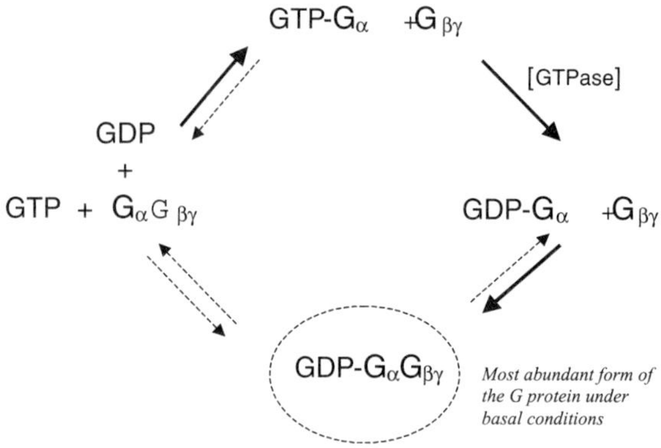

FIGURE 2.7 G protein activation cycle under basal conditions. The solid arrows show the fast reactions, whereas the dashed arrows show the slow or rate-limiting steps. The dashed ellipse encircles the most prominent form of the G protein under basal conditions.

In order to model the entire system, that is, to determine how occupation of the receptor by an agonist influences the rate of formation of GTP-G_α, we need to take into account the steady-state condition. A comprehensive analysis of this type has been described by Waelbroeck et al.[33] The steady-state condition can be modeled through a combination of equilibrium and kinetic methods. The rapidly equilibrating binding steps can be modeled as equilibria, whereas the slow binding steps (e.g., dissociation of GDP from the G protein) and the GTPase activity require a kinetic component (differential equations). In addition, the concentration of GTP in the cytosol is maintained by kinase activity that converts GDP to GTP in the cytosol. This step also needs to be addressed in the kinetics. From the perspective of ligand binding to the receptor, we need to consider whether the GTPase-mediated conversion of GTP-G_α to GDP-G_α alters ligand binding. Since several studies have shown that GTP and GDP have similar effects on the binding of agonists to G protein-linked receptors, then it would appear that the GTPase activity does not alter ligand binding.[34,35] Thus, in many instances, it should be appropriate to use an equilibrium analysis to model ligand binding to the receptor in the presence of GTP.

A model for the interaction of guanine nucleotide with a G protein-linked receptor is shown in Figure 2.8. This model was derived by taking the simple ternary complex model and adding a guanine nucleotide binding step.[36] The bottom side of the cube represents the simple ternary complex model, whereas the top side shows the same equilibria with a guanine nucleotide (X) bound to the G protein (i.e., GX). Each vertical edge of the cube corresponds to the binding of X to a different G protein complex (i.e., G, RG, and DRG), with the exception of the two rear edges which are the same. The microscopic affinity constants K_1, K_2, and K_3 describe the binding of drug to receptor, the binding of G protein to receptor, and the binding of guanine nucleotide to the G protein, respectively. The Greek letters are used to

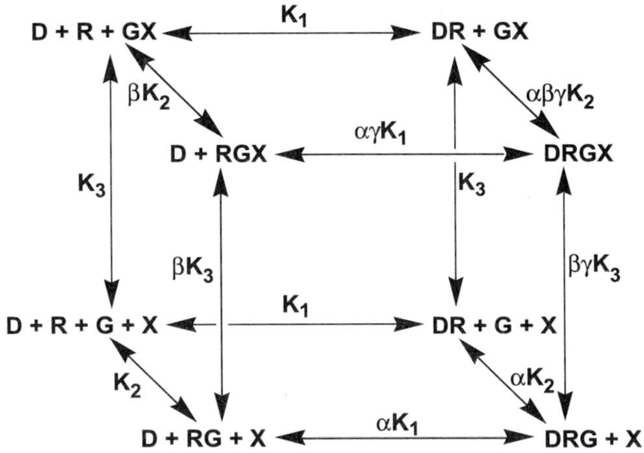

FIGURE 2.8 Ternary complex model with guanine nucleotide (X). The model shows the equilibrium between a drug (D), receptor (R), G protein (G), and guanine nucleotide (X). The model is described at the level of receptor complexes and their microscopic affinity constants. Each receptor complex shown in the model (e.g., DR) actually represents an equilibrium between two receptor states (e.g., $DR_s \leftrightarrow DR_s{}^*$) not shown in the figure. The definition of the variables and equilibrium constants in the model are described in the text and Appendix of this chapter. It is possible to define the microscopic constants for the receptor complexes shown in the model in terms of the more fundamental microscopic constants of the receptor states as described in the Appendix of this chapter (see Equations A49–A61).

denote cooperativity. α denotes the cooperativity between the binding of D and G to R, β denotes the cooperativity between the binding of R and X to G, and γ denotes the cooperativity between the binding of D and X to the RG complex. The other cooperativities are factors of the latter three. With the exception of the rear face, all other faces of the cube exhibit allosteric interactions. The solution to this cube of equilibria has been described previously[36] and is also given in the Appendix.

The guanine nucleotide is indicated in Figure 2.8 as X because it can represent GTP, GDP, or their analogs. However, with regard to GTP, the GX complex is mainly in the form of GTP-G_α, whereas the RGX complex is mainly in the form of RG-GDP (see discussion above). Thus, the model does not adequately address the form of the guanine nucleotide as far as GTP is concerned; nevertheless, it should accurately describe the influence of GTP and GDP on agonist binding to the receptor.

In examining the behavior of the model in Figure 2.8 with respect to different agonists, it is essential to rely on conformational analysis. Agonists differ with respect to their microscopic affinity constants for the ground and active conformations of the receptor. These microscopic constants determine the values of the cooperativity constants α and γ in the model shown in Figure 2.8. Thus, α and γ do not vary independently for different agonists; rather, they exhibit an inverse relationship that depends on the ratio of microscopic affinity constants of the ground and active conformations of the receptor (a). This relationship is described in the Appendix and is illustrated in Figure 2.9. In the following examples (i.e., Figures

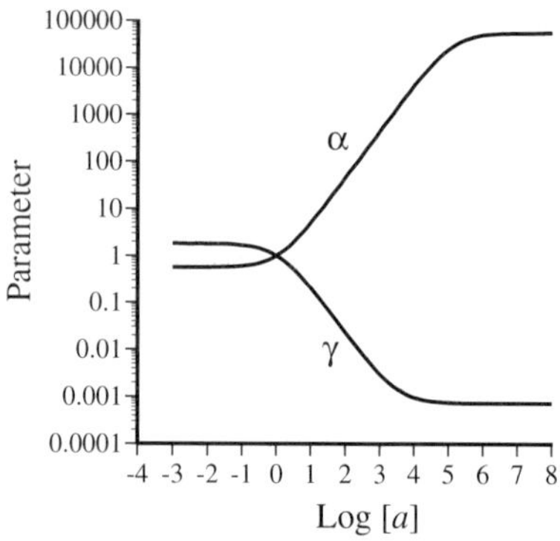

FIGURE 2.9 The effect of the ratio of microscopic affinity constants (i.e., aK_d/K_a or simply a) of a drug for the ground (K_a) and active (aK_a) conformations of the receptor on the cooperativity between the binding of the drug and G protein to the receptor (α) and on the cooperativity between the binding of the drug and guanine nucleotide to the receptor-G protein complex (γ). The values for α and γ where calculated as a function of a as described in Equations A59 and A61 in the Appendix, respectively. For this example, the following values for the microscopic constants of the receptor conformations were used: $K_a = 1.6 \times 10^5$ M^{-1}, $K_c = 8 \times 10^{-6}$, $b = 10^5$, and $c = 4 \times 10^{-4}$.

10 through 13), the values of the parameters were constrained to those allowed by conformational analysis (see Appendix, Equations A49 through A61).

Figure 2.10 shows the behavior of the ternary complex model with guanine nucleotide when there is an excess of G protein to receptor. Theoretical curves are shown for the binding of a highly efficacious agonist to a receptor that lacks constitutive activity (i.e., $K_2 << 1$). Figure 2.10A shows receptor occupancy in the presence of increasing concentrations of guanine nucleotide. It can be seen that binding is consistent with a simple one-site model, and that the guanine nucleotide causes a reduction in the apparent affinity of the agonist for the receptor. The effect of the guanine nucleotide reaches a plateau at high concentrations. The maximal shift in the occupancy curve caused by the guanine nucleotide (guanine nucleotide shift) is equivalent to 1600-fold. Figure 2.10B shows the effects of guanine nucleotide on the amount of drug bound in the form of quaternary complex (*DRGX*). Guanine nucleotide causes a concentration-dependent reduction in the observed affinity of the quaternary complex and an increase in the maximal amount formed at high agonist concentration. Figure 2.10C shows the amount of bound agonist in the form of ternary complex (*DRG*). It can be seen that the guanine nucleotide reduces both the observed affinity and maximal amount of ternary complex. These results are summarized in Table 2.7. In the presence of a given concentration of guanine nucleotide, the observed affinities of the various complexes (i.e., *DR*, *DRG*, *DRGX*) are all the same as shown by the similarity in the observed dissociation constants of the three complexes (see Table 2.7). The behavior of the model in Figure 2.10 is similar when a less efficacious agonist is used, except that both the maximum guanine nucleotide-shift and the maximal amount of *DRGX* complex formed are less.

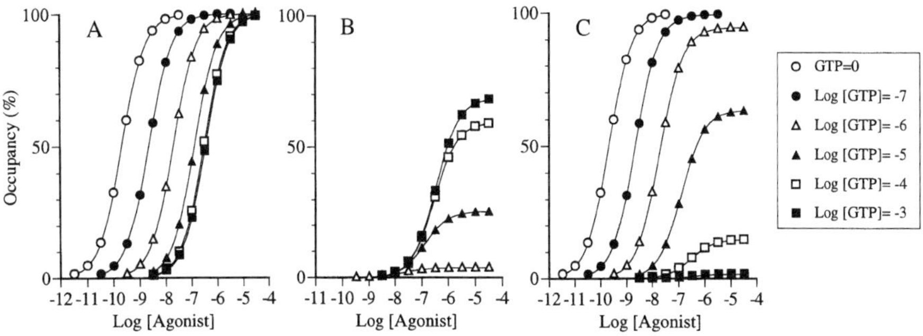

FIGURE 2.10 The influence of guanine nucleotide (GTP) on the binding of an agonist to a receptor that conforms to the ternary complex model with excess G protein ($G_T/R_T = 10$). A. Total receptor occupancy (i.e., $DRGX + DRG + DR$). B. Occupancy of the quaternary complex ($DRGX$). C. Occupancy of the ternary complex (DRG). For this example, the following microscopic constants for the receptor conformations were used: $a = 10^6$, $b = 10^5$, $c = 4 \times 10^{-4}$, $K_a = 10^5$ M^{-1}, $K_b = 7 \times 10^{-3}$, $K_c = 8 \times 10^{-6}$, and $K_d = 10^8$ M^{-1}. These microscopic constants determine the microscopic constants for the various receptor complexes: $\alpha = 4.94 \times 10^4$, $\beta = 0.556$, $\gamma = 7.22 \times 10^{-4}$, $K_1 = 9 \times 10^5$ M^{-1}, $K_2 = 1.26 \times 10^{-2}$, and $K_3 = 10^8$ M^{-1}. The observed binding parameters for this example are given in Table 2.7.

TABLE 2.7
Effects of Guanine Nucleotide (e.g., GDP or GTP) on the Binding of an Agonist to a Receptor That Conforms to the Ternary Complex Model with Excess G Protein ($G_T/R_T = 10$). The binding parameters were estimated from the data shown in Figure 2.10.

	Guanine Nucleotide (M)						
	0	10^{-8}	10^{-7}	10^{-6}	10^{-5}	10^{-4}	10^{-3}
			Receptor-G Protein Complex				
pK_{obs}[a]	9.67	9.39	8.66	7.72	6.90	6.54	6.47
$DRGX_{max}$[b]	0	0.04%	0.40%	3.79%	25.4%	59.5%	68.7%
DRG_{max}[c]	99.98%	99.9%	99.4%	94.5%	63.4%	14.8%	1.71%
			Receptor Complex				
pK_{obs}[d]	9.63	9.34	8.61	7.67	6.86	6.50	6.44
DR_{max}[e]	0.02%	0.04%	0.20%	1.70%	11.2%	25.7%	29.6%

[a] Negative logarithm of the concentration of agonist required for half-maximal occupancy of the DRGX and DRG complexes.
[b] Maximal amount of DRGX complex formed at 100% receptor occupancy.
[c] Maximal amount of DRG complex formed at 100% receptor occupancy.
[d] Negative logarithm of the concentration of agonist required for half-maximal occupancy of the DR complex.
[e] Maximal amount of DR complex formed at 100% receptor occupancy.

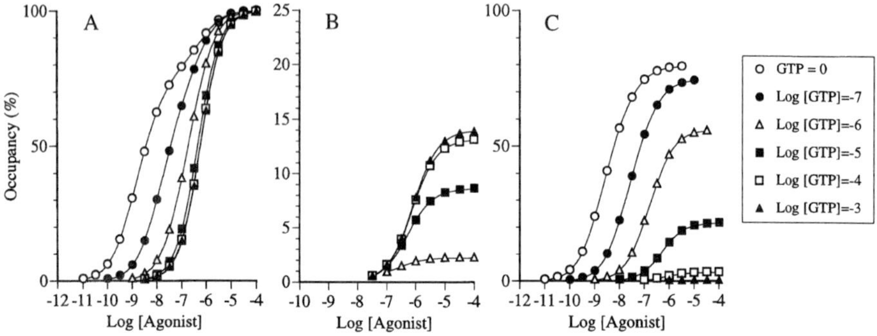

FIGURE 2.11 The influence of guanine nucleotide (GTP) on the binding of an agonist to a receptor that conforms to the ternary complex model with limiting G protein ($G_T/R_T = 0.8$). A. Total receptor occupancy (i.e., *DRGX* + *DRG* + *DR*). B. Occupancy of the quaternary complex (*DRGX*). C. Occupancy of the ternary complex (*DRG*). For this example, the following microscopic constants for the receptor conformations were used: $a = 10^6$, $b = 10^5$, $c = 4 \times 10^{-4}$, $K_a = 1.6 \times 10^5$ M^{-1}, $K_b = 7 \times 10^{-3}$, $K_c = 8 \times 10^{-6}$, and $K_d = 10^8$ M^{-1}. These microscopic constants determine the microscopic constants for the various receptor complexes: $\alpha = 4.94 \times 10^4$, $\beta = 0.556$, $\gamma = 7.22 \times 10^{-4}$, $K_1 = 1.44 \times 10^6$ M^{-1}, $K_2 = 1.26 \times 10^{-2}$, and $K_3 = 10^8$ M^{-1}. The observed binding parameters for this example are given in Table 2.8.

Figure 2.11 shows the behavior of the ternary complex model with guanine nucleotide when there is limiting G protein ($G_T/R_T = 0.8$). Theoretical curves are shown for the binding of a highly efficacious agonist to a receptor that lacks constitutive activity (i.e., $K_2 << 1$). Figure 2.11A shows receptor occupancy in the presence of increasing concentrations of guanine nucleotide. It can be seen that binding is consistent with a two-site model in the absence of guanine nucleotide, and that the guanine nucleotide causes a reduction in both the observed affinity and the proportion of the high affinity component. Conversely, guanine nucleotide increased the low affinity component without influencing its observed affinity. Figure 2.11B shows the effects of guanine nucleotide on the amount of drug bound in the form of quaternary complex (*DRGX*). Guanine nucleotide causes a concentration-dependent reduction in the observed affinity of the quaternary complex and an increase in the maximal amount formed. Figure 2.11C shows the amount of agonist bound in the form of ternary complex (*DRG*). It can be seen that the guanine nucleotide reduces both the observed affinity and maximal amount of ternary complex. These results are summarized in Table 2.8. It can be seen that the observed affinities of the *DRG* and *DRGX* complexes are the same at the various concentrations of guanine nucleotide. Together, these two complexes combine to form the high affinity component of the agonist binding curve, whereas the *DR* complex forms the low affinity component. Thus, increasing concentrations of guanine nucleotide convert the high affinity ternary complex (*DRG*) into a low affinity quaternary complex (*DRGX*). Moreover, the maximal amount of *DRGX* complex formed in the presence of guanine nucleotide is less than that of the ternary complex in the absence of guanine nucleotide. The behavior of the model in Figure 2.11 is similar when a less efficacious agonist is used except that both the maximum

TABLE 2.8
Effects of Guanine Nucleotide (e.g., GDP or GTP) on the Binding of an Agonist to a Receptor That Conforms to the Ternary Complex Model with Limiting G Protein (G_T/R_T = 0.8). The binding parameters were estimated from the data shown in Figure 2.11.

	Guanine Nucleotide (M)						
	0	10^{-8}	10^{-7}	10^{-6}	10^{-5}	10^{-4}	10^{-3}
			Receptor-G Protein Complex				
pK_{obs}[a]	8.71	8.42	7.74	6.99	6.49	6.32	6.30
$DRGX_{max}$[b]	0	0.03%	0.30%	2.25%	8.66%	13.2%	14.0%
DRG_{max}[c]	79.4%	78.8%	74.5%	56.0%	21.6%	3.29%	0.35%
			Receptor Complex				
pK_{obs}[d]	6.19	6.21	6.30	6.37	6.28	6.22	6.22
DR_{max}[e]	20.6%	21.2%	25.2%	41.8%	69.8%	83.5%	85.6%

[a] Negative logarithm of the concentration of agonist required for half-maximal occupancy of the DRGX and DRG complexes.
[b] Maximal amount of DRGX complex formed at 100% receptor occupancy.
[c] Maximal amount of DRG complex formed at 100% receptor occupancy.
[d] Negative logarithm of the concentration of agonist required for half-maximal occupancy of the DR complex.
[e] Maximal amount of DR complex formed at 100% receptor occupancy.

guanine nucleotide-shift and the maximal amounts of *DRG* and *DRGX* complexes formed are less.

Figure 2.12 shows the effects of a high concentration of GTP on the binding curves of a highly efficacious agonist (panel A), a less efficacious agonist (panel B), an antagonist (panel C) and an intrinsic antagonist (panel D). In this example, the receptor lacks constitutive activity, and there is limiting G protein (G_T/R_T = 0.8). It can be seen that the GTP-induced reduction in the observed affinities of the agonists is correlated with intrinsic efficacy and that GTP has no effect on the binding of antagonists. A similar example is shown in Figure 2.13 for a constitutively active receptor. In this case, GTP reduces the observed affinity of the agonists (see Figure 2.13A and B), although this effect is less than that observed for the receptor lacking constitutive activity shown in Figure 2.12. GTP has no effect on the binding of the antagonist to the constitutively active receptor (Figure 2.13C), whereas GTP increases the observed affinity of the intrinsic antagonist for the constitutively active receptor (Figure 2.13D). In this case, GTP converts the two-site binding curve of the intrinsic antagonist into a simple one-site curve of high affinity.

The examples in Figures 2.12 and 2.13 show that the inhibitory effect of GTP on the observed affinity of agonists (GTP-shift) is correlated with the ratio of affinity constants of the agonist for the ground and active conformations of the receptor. This relationship implies that the GTP-shift should be proportional to the intrinsic efficacy of the agonist. Such a relationship was observed in a prior study on cardiac

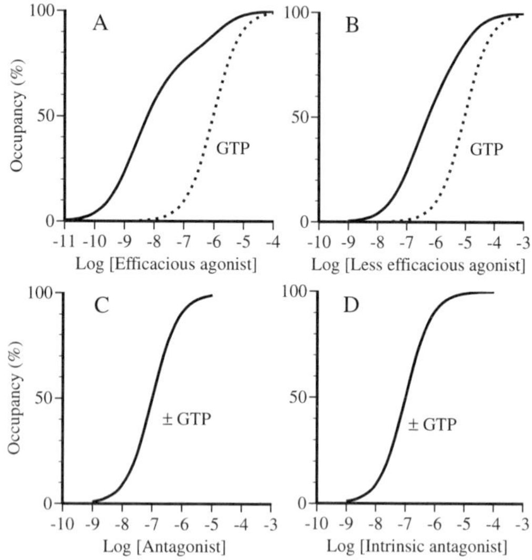

FIGURE 2.12 The influence of guanine nucleotide (GTP) on the binding of drugs to a receptor that conforms to the ternary complex model with limiting G protein ($G_T/R_T = 0.8$) and little constitutive activity. The curves represent total receptor occupancy ($DRGX + DRG + DR$) in the absence (solid lines) and presence (dotted lines) of guanine nucleotide. The microscopic state constants for this receptor-G protein system are: $K_b = 7 \times 10^{-3}$; $K_c = 8 \times 10^{-6}$, $K_d = 10^8$ M^{-1}, $b = 10^5$, and $c = 4 \times 10^{-4}$. These microscopic state constants determine the microscopic constant for the receptor-G protein complex ($K_2 = 1.26 \times 10^{-2}$) and the cooperativity between the binding of guanine nucleotide and receptor with the G protein ($\beta = 0.556$). The ligand-dependent microscopic constants are given in connection with each panel of the figure as follows. A. The binding of an efficacious agonist. The microscopic state constants are: $K_a = 10^5$ M^{-1} and $a = 10^6$. The microscopic constants for the ligand–receptor complexes are: $K_1 = 9 \times 10^5$ M^{-1}, $\alpha = 4.94 \times 10^4$, and $\gamma = 7.22 \times 10^{-4}$. B. The binding of a less efficacious agonist. The microscopic state constants are: $K_a = 10^5$ M^{-1} and $a = 10^4$. The microscopic constants for the ligand–receptor complexes are: $K_1 = 1.08 \times 10^5$ M^{-1}, $\alpha = 4.12 \times 10^3$, and $\gamma = 9.45 \times 10^{-4}$. C. The binding of an antagonist. The microscopic state constants are: $K_a = 10^7$ M^{-1} and $a = 1.0$. The microscopic constants for the ligand–receptor complexes are: $K_1 = 10^7$ M^{-1}, $\alpha = 1.0$, and $\gamma = 1.0$. D. The binding of an intrinsic antagonist. The microscopic state constants are: $K_a = 10^7$ M^{-1} and $a = 10^{-4}$. The microscopic constants for the ligand–receptor complexes are: $K_1 = 10^7$ M^{-1}, $\alpha = 0.556$, and $\gamma = 1.8$.

M_2 receptors, where the GTP-shift was shown to be highly correlated with the relative efficacy of agonists as estimated from their ability to mediate an inhibition of adenylyl cyclase activity (see Figure 2.14).[20]

2.3.3 Effects of Agonists on the Binding of Guanine Nucleotides

By solving the model of Figure 2.8 for the binding of guanine nucleotide (X) instead of D, it is possible to describe the effects of agonists on the binding of guanine nucleotides (see Appendix). However, when used to measure GTP binding, the model

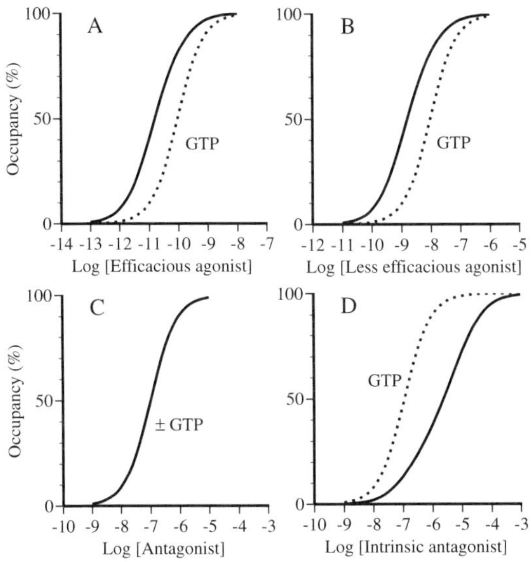

FIGURE 2.13 The influence of guanine nucleotide (GTP) on the binding of drugs to a constitutively active receptor that conforms to the ternary complex model with limiting G protein ($G_T/R_T = 0.8$). The curves represent total receptor occupancy ($DRGX + DRG + DR$) in the absence (solid lines) and presence (dotted lines) of guanine nucleotide. The microscopic state constants for this receptor-G protein system are: $K_b = 7 \times 10^{-3}$, $K_c = 0.1$, $K_d = 10^8$ M^{-1}, $b = 10^5$, and $c = 4 \times 10^{-4}$. These microscopic state constants determine the microscopic constant for the receptor-G protein complex ($K_2 = 63.6$) and the cooperativity between the binding of guanine nucleotide and receptor with the G protein ($\beta = 5 \times 10^{-4}$). The ligand-dependent microscopic constants are given in connection with each panel of the figure as follows. A. The binding of an efficacious agonist. The microscopic state constants are: $K_a = 10^5$ M^{-1} and $a = 10^6$. The microscopic constants for the ligand–receptor complexes are: $K_1 = 9.09 \times 10^9$ M^{-1}, $\alpha = 11$, and $\gamma = 0.8$. B. The binding of a less efficacious agonist. The microscopic state constants are: $K_a = 10^5$ M^{-1} and $a = 10^4$. The microscopic constants for the ligand–receptor complexes are: $K_1 = 9.1 \times 10^7$ M^{-1}, $\alpha = 11$, and $\gamma = 0.8$. C. The binding of an antagonist. The microscopic state constants are: $K_a = 10^7$ M^{-1} and $a = 1.0$. The microscopic constants for the ligand–receptor complexes are: $K_1 = 10^7$ M^{-1}, $\alpha = 1.0$, and $\gamma = 1.0$. D. The binding of an intrinsic antagonist. The microscopic state constants are: $K_a = 10^7$ M^{-1} and $a = 10^{-4}$. The microscopic constants for the ligand–receptor complexes are: $K_1 = 9.09 \times 10^6$ M^{-1}, $\alpha = 2.22 \times 10^{-4}$, and $\gamma = 10^3$.

may be somewhat erroneous because of the steady-state issues raised above in connection with GTPase activity. Nevertheless, the model should be appropriate for the analysis of the binding of GDP as well as nonhydrolyzable guanine nucleotides (e.g., GTPγS).

Figure 2.15 shows the effects of agonists on the binding of GDP to a G protein that interacts with a receptor lacking constitutive activity (i.e., $K_2 << 1$). The theoretical curves conform to the ternary complex model with excess G protein. Figure 2.15A shows GDP binding curves in the presence of a high concentration of an efficacious agonist. The parameters used here are identical to those used in Figure

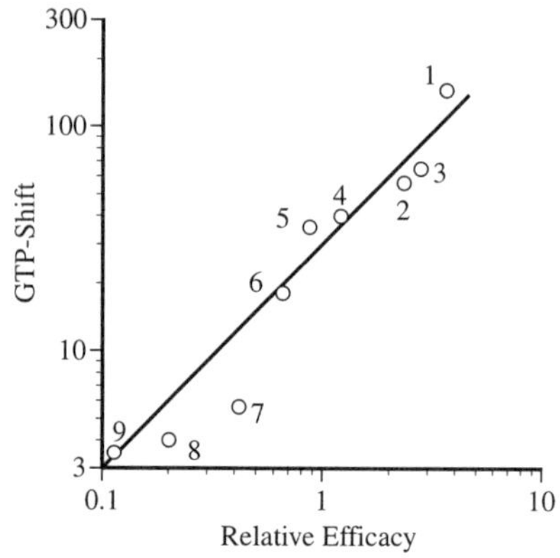

FIGURE 2.14 The relationship between the inhibitory effects of GTP on agonist binding to M_2 muscarinic receptors and the relative efficacy of the agonist. The concentration of agonist required for half-maximal receptor occupancy in the presence of GTP (0.1 mM) divided by that measured in the absence of GTP is plotted against the relative efficacy of the agonist. Receptor occupancy was determined in muscarinic receptor binding experiments on the rabbit myocardium, and relative efficacy was estimated in adenylyl cyclase assays. The data are from Ehlert.[20] 1, oxotremorine-M; 2, carbachol; 3, *cis*-dioxolane; 4, oxotremorine; 5, (+) aceclidine; 6, (–) aceclidine; 7, N-methylaceclidine; 8, BM5; and 9, BOK1.

2.10, however, in this instance, the binding of the GDP is plotted instead of that of the agonist. It can be seen that GDP binding is consistent with a simple one-site model, and that the agonist causes a reduction in the apparent affinity of GDP. However, the shift (1.5-fold) in the GDP binding curve is much less than that caused by the guanine nucleotide on the agonist binding curve (1600-fold; see Figure 2.10). This result may seem to violate the laws of thermodynamics that require D and X to have reciprocal allosteric effects at the level of the $DRGX$ complex. However, because there is an excess of G protein to receptor, the net effect of D on the binding of X to all of the G proteins is less than the effect of X on the binding of D to a much smaller number of receptors. Accordingly, when the ratio of G_T to R_T is approximately the same (i.e., 0.8), then the effect of the agonist on the binding of X is much greater (see Figure 2.15B). Figure 2.15C shows the binding of GDP in the presence of a high concentration of different agonists with varying values of a (i.e., the ratio of microscopic affinity constants for the ground (K_a) and active (aK_a) states of the receptor). Over a range of values, a is proportional to intrinsic efficacy as described in a later section. It can be seen that as the value of a decreases, the effect of the agonist on GDP binding also decreases. Since a is correlated with intrinsic efficacy, the model shows that the ability of the agonist to reduce the observed affinity of GDP is proportional to intrinsic efficacy. As described above, the conformational constraints of the model require a reciprocal relationship between α and γ. Thus, as α increases, due to an increase in a, there is a corresponding decrease in γ (i.e., more negative cooperativity between the binding of D and X to the RG complex).

Figure 2.15D shows the effects of a high concentration of an intrinsic antagonist ($a = 10^{-4}$) on the binding of GDP to a G protein in equilibrium with a constitutively

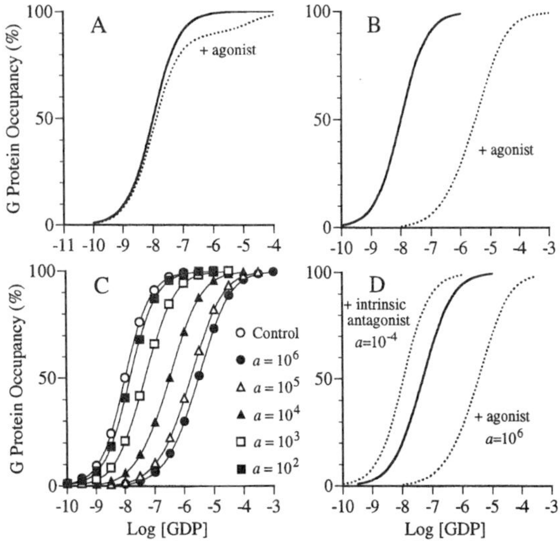

FIGURE 2.15 Effects of drugs on the binding of GDP to G protein. In each example, the concentration of the drug (i.e., agonist or antagonist) was set at 10^{-3} M. A. The effects of a highly efficacious agonist on the binding of GDP to a G protein in 10-fold excess relative to the receptor (i.e., $G_T/R_T = 10$), which has little constitutive activity. The microscopic constants for this example are identical to those of Figure 2.9. B. The effects of a highly efficacious agonist on the binding of GDP to a G protein with an abundance of 80% (i.e., $G_T/R_T = 0.8$) relative to the receptor, which has little constitutive activity. The microscopic constants for the receptor conformations are identical to those of Figure 2.9. C. The effects of agonists with varying intrinsic efficacies (i.e., varying a values) on the binding of GDP to a G protein with an abundance of 80% (i.e., $G_T/R_T = 0.8$) relative to the receptor, which has little constitutive activity. The microscopic constants for the receptor conformations are identical to those of Figure 2.9, with the exception of the ratio of microscopic affinity constants for the ground and active conformations of the receptor (a), which was varied as indicated in panel C. D. The effects of an efficacious agonist ($a = 10^6$) and an intrinsic antagonist ($a = 10^{-4}$) on the binding of GDP to a G protein with an abundance of 80% (i.e., $G_T/R_T = 0.8$) relative to the receptor, which is constitutively active. The microscopic constants for the receptor conformations are identical to those of Figure 2.9, with the exception of K_c (8×10^{-3}) and the ratio of microscopic affinity constants for the ground and active conformations of the receptor (a), which was varied as indicated in panel D.

active receptor. It can be seen that the intrinsic antagonist actually increases the observed affinity of GDP for the G protein. This action would inhibit the exchange of GTP for GDP at the G protein and cause an inhibition of spontaneous basal activity. For comparison, the effects of an efficacious agonist are also shown in Figure 2.15D.

A common method for investigating the function of G protein-linked receptors is to measure agonist-stimulated [^{35}S]GTPγS binding.[37] Typically, agonists have little effect on the binding of [^{35}S]GTPγS alone, but cause a modest increase when measured in the presence of GDP. Thus, agonists cause a reduction in the affinity of GDP, which is manifested indirectly as an increase in [^{35}S]GTPγS binding due to the diminished competitive effect of GDP. The modeling shown in Figure 2.15 is consistent with this mechanism of agonist-mediated reduction in the observed affinity of GDP. Since this effect of agonists is often small and sometimes difficult to detect, then G proteins probably outnumber receptors at least several fold. Such a stoichiometry could account for a small agonist effect as indicated by a comparison of the conditions shown in Figures 2.15A and B. This model also predicts that agonists should reduce the affinity of [^{35}S]GTPγS as well; however, typically little direct effect on [^{35}S]GTPγS binding has been observed. This situation might be explained by the preferential binding of [^{35}S]GTPγS to free G_α subunits, which do not directly communicate with the receptor unless they complex with $G_{\beta\gamma}$. Also, a simple excess of G_α to $G_{\beta\gamma}$ could further isolate the [^{35}S]GTPγS-G_α complex from the receptor system. Finally, the routine use of low concentrations of [^{35}S]GTPγS also tends to favor the [^{35}S]GTPγS-G_α complex.

Actually, an agonist-mediated reduction in the affinity of both GDP and GTP for the *DRGX* complex is the most mechanistically sound means for an agonist to enhance signaling through a G protein-linked receptor.[38] A reduction in the affinity of GTP would not cause a decrease in the binding of GTP under physiological conditions, because GTP is maintained at near millimolar concentrations in the cytosol, whereas its affinity for G_α is in the nanomolar range. Also, kinases in the cytoplasm continuously convert GDP to GTP, making the concentration of GDP in the cytosol much lower than that of GTP. Thus, at equilibrium, GTP would easily "out-compete" GDP and saturate G proteins. However, the GTPase activity of G_α is much faster than the dissociation rate of GDP, giving rise to a steady-state in which G proteins are saturated with GDP in the basal state (see discussion above). By increasing the dissociation rate of GDP, the agonist receptor complex drives the system closer to a thermodynamic equilibrium in which GTP rapidly replaces GDP in the G protein. An advantage to having the agonist reduce the affinity of GTP for the G protein is that this mechanism tends to destabilize the $DRGX_T$ complex, enabling the GTP-G_α complex to rapidly leave the receptor and activate its effector. This situation would increase the turnover rate of the system and lead to a greater rate of production of GTP-G_α subunits. In contrast, an agonist-mediated increase in the affinity of GTP would stabilize the $DRGX_T$ complex, thus hindering its rapid dissociation and decreasing the concentration of GTP-G_α subunits in the cytosol.

In much of the foregoing discussion, the mechanism of G protein-linked receptor activation has been described as an agonist-mediated decrease in the observed affinity of GDP for the G protein. This focus on affinity is due in large part to the nature of the analysis, which is primarily an equilibrium approach. However, the agonist-mediated reduction in the affinity of GDP is not the essential mechanism of G protein activation. The essential step is an increase in the dissociation kinetics of GDP. Admittedly, this kinetic change would most likely be manifested as reduced affinity at equilibrium; however, it does not have to be. Many G_q-linked receptors do not exhibit detectable GTP-shifts for agonist binding.[39-42] Also, the agonist binding curves in the absence of GTP often resemble a one-site model of low affinity with little evidence of a high affinity component representing the ternary complex.[39-42] These observations may be due, in part, to the lower expression of G_q compared to G_i.[15] Alternatively, agonists may not cause much of a reduction in the affinity of

GDP for G_q. Nevertheless, agonists could mediate an increase in the dissociation kinetics of GDP with no change in affinity provided that the increased off rate is opposed by a corresponding increase in association rate.

This mechanism may not seem so unlikely when one considers the possibility that, in the inactive state, the G protein may engulf GDP. That is, the GDP may be locked within a binding pocket so that it cannot dissociate or associate. Activation of the G protein may correspond to an opening up of the binding pocket, thus enabling both a rapid association and dissociation.

2.4. RECEPTOR THEORY AND THE TERNARY COMPLEX MODEL

It is ironic that scientists universally employ the terms affinity and efficacy to describe drug-receptor interactions, yet a consensus on the precise definition of these terms is lacking. The terms are used intuitively to refer to the potency and functional activity of the drug-receptor complex. However, on a more philosophical level, the terms are often regarded as bewildering concepts that require further development in technology to crack their mystery. This view often leads to the conclusion that classic receptor theory must be flawed and cannot possibly account for the behavior of G protein-linked receptors.[43]

Part of the conflict arises from a discrepancy between what is commonly perceived to be the original intent of the authors of receptor theory and the mathematics of the theory itself. It follows that if the purpose of a method is one thing, yet the method actually measures something else, then the method is flawed by design. However, if the mathematical theory can be shown to be extremely powerful for measuring another aspect of receptor properties, should it not be valid for that purpose? We might even be inclined to state that the purpose of the theory is to measure this other aspect.

Therefore it is necessary to consider first some commonly held views about affinity and intrinsic efficacy and to discuss the extent to which they are consistent with the mathematics of receptor theory. Second, and more significantly, it is important to show how the mathematics of receptor theory apply to G protein-linked receptors.

2.4.1 Traditional View of Receptor Theory

Affinity and intrinsic efficacy are traditionally viewed as independent parameters — affinity being a measure of the agonist's ability to bind to the receptor in the first place and intrinsic efficacy being a measure of the agonist's ability to activate the receptor or to induce the active conformation.[44] However, the mathematics of receptor theory are not so constrained. They are based on the assumption that the stimulus (i.e., the degree of receptor activation) is equivalent to the product of receptor occupancy and intrinsic efficacy.[45,46] Receptor occupancy is governed by the observed dissociation constant (see Tables). However, the observed dissociation constant (K_{obs}) is a function of the microscopic affinity constants for the ground and active conformations of the receptor (i.e., aK_a and K_a) as well as the ability of the receptor to undergo the conformational change (K_c). Intrinsic efficacy is also a function of these constants. Over a range of values, intrinsic efficacy is proportional to the ratio, aK_a/K_a. Thus, affinity and intrinsic efficacy are functions of the same parameters, which means they are not independent. For example, we could rewrite the mathematics of receptor theory using the function $f(aK_a, K_a)$ for observed affinity and the function $g(aK_a, K_a)$ for intrinsic efficacy. The mathematics would be equally valid.

The traditional view of affinity is that this parameter is a property of the shape of the ground state of the receptor, and therefore, it should be represented quantitatively by the microscopic affinity constant, K_a. In the ternary complex model, K_1 is approximately equal to K_a when the receptor lacks constitutive activity. Although it is possible to rewrite the mathematics of receptor theory in terms of K_a and aK_a as just described, it is usually not possible to measure these parameters easily. Furchgott's method[46,47] of partial receptor inactivation, which embraces the mathematics of classic receptor theory, provides an estimate of the observed K_D, not K_a. Nevertheless, the observed K_D is an extremely useful parameter in its own right. Given the recent debate over the validity of classic receptor theory, it is ironic that the reciprocal of the observed K_D for a G protein-linked receptor is probably not much different from K_a or K_1 when the receptor is functioning under normal physiological conditions (i.e., high concentrations of GTP).

Another traditional view of receptor theory is that affinity and intrinsic efficacy are intrinsic properties of the receptor in isolation. This view readily applies to ligand-gated ion channels because these proteins are both receptors and effectors. They do not depend on their association with other proteins to trigger a cellular signal; consequently, the electrophysiologist can measure the current through the channel as a direct measure of receptor activation. However, G protein-linked receptors depend on their association with G proteins to trigger cellular responses. Likewise, scientists are usually dependent on the receptor–G protein interaction to measure the activity of the agonist-receptor complex. Consequently, the observed measurements of affinity and receptor activation are distorted by the G protein, and these estimates do not necessarily reflect those of the receptor in isolation. Similarly, classic pharmacological null methods were designed to enable scientists to measure receptor properties in their native context. Consequently, these methods yield estimates of observed affinity and relative efficacy that are influenced by G proteins.

2.4.2 The Mathematics of Receptor Theory and the Ternary Complex Model

The mathematics of receptor theory considers the response to an agonist as a train of events. The initial events are receptor-dependent and are defined as the stimulus, whereas the latter events are independent of the receptor and are conveniently known as the stimulus–response relationship. The stimulus is a function that defines the relationship between receptor activation and the agonist concentration. Thus, to assess whether classic receptor theory is consistent with the ternary complex model, we need to determine whether the ternary complex model is consistent with the stimulus function.

In its essence, the stimulus (S) is a measure of receptor activation expressed as a function of the agonist concentration (D).[45,46] It is defined as the product of receptor occupancy (P) and intrinsic efficacy (ε):

$$S = \varepsilon \cdot P \tag{14}$$

Receptor occupancy is given by the equation:

$$P = \frac{D \cdot R_T}{D + K_{obs}} \tag{15}$$

in which R_T denotes the total receptor concentration and K_{obs} denotes the observed dissociation constant of the agonist. Substituting Equation (15) for receptor occupancy in Equation (14) yields:

$$S = \frac{\varepsilon \cdot D \cdot R_T}{D + K_{obs}} \tag{16}$$

Thus, to determine if the mathematics of receptor theory are consistent with the ternary complex model, we need to determine whether the model is consistent with Equation (16).

Before making this determination, we need to address the units of intrinsic efficacy. In practice, it is usually not possible to obtain absolute estimates of R_T and intrinsic efficacy from functional data; nevertheless, it is possible to estimate the intrinsic efficacy of an agonist relative to that of another agonist. Consequently, we are at some liberty to assign a specific value of intrinsic efficacy to an agonist as long as it is appropriate relative to those of other agonists. Classic receptor theory considers the stimulus as a unit-less quantity; consequently, Furchgott[46] assigned intrinsic efficacy units of the reciprocal of the concentration of receptors so that the product of intrinsic efficacy and receptor occupancy would be unit-less. However, it would seem more appropriate to assign intrinsic efficacy units of active receptor complex per total receptor concentration so that the stimulus yields a measure of the number of active receptor complexes. Practically speaking, this means that intrinsic efficacy should be unit-less, because the units of the active receptor complex and total receptor complex are the same.

Within the context of G protein-linked receptors, receptor activation corresponds to the rate at which the agonist-receptor complex generates G_α-GTP complexes in the cytosol. Since the dissociation of GDP is the rate-limiting step in this process, then receptor activation is proportional to the amount of G-GDP complexes that the agonist destabilizes at any given point in time. This amount is simply proportional to the amount of the active receptor conformation in the quaternary complex ($DR_s{*}GX$). Thus, the stimulus function and the function for $DR_s{*}GX$ should be equivalent, if receptor theory is consistent with the ternary complex model. Figures 2.16A and B show the behavior of the ternary complex model in the presence of

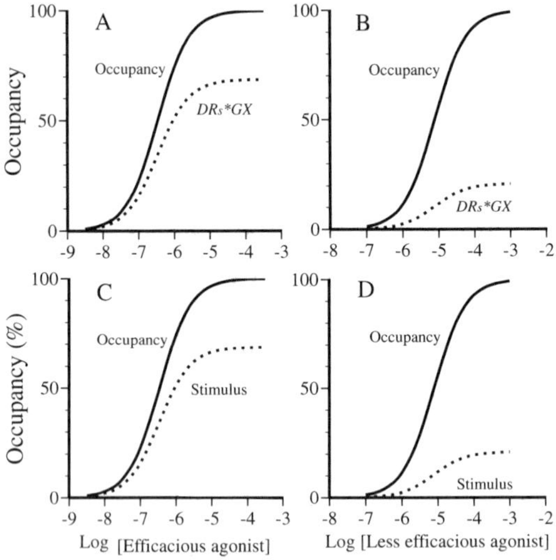

FIGURE 2.16 Comparison between classic receptor theory and the ternary complex model for a receptor with excess G protein (G_T/R_T = 10) and little constitutive activity. In each example, the concentration of guanine nucleotide was set at 10^{-3} M. A. Comparison between total receptor occupancy (*DRGX* + *DRG* + *DR*) and drug bound in the form of active quaternary complex (DR_s*GX) for an efficacious agonist (a = 10^6). The microscopic constants for the receptor conformations in this example are identical to those of Figure 2.9. The observed dissociation constant of the agonist (K_{obs}) is approximately 3.39 × 10^{-7} M. B. Comparison between total receptor occupancy (*DRGX* + *DRG* + *DR*) and drug bound in the form of active quaternary complex (DR_s*GX) for a less efficacious agonist (a = 10^4). The microscopic constants for the receptor conformations in this example are identical to those of Figure 2.9 with the exception of a, which was set at 10^4. The K_{obs} of the agonist is approximately 7.76 × 10^{-6} M. C. Comparison between receptor occupancy and the stimulus function for an efficacious agonist (ε = 0.69). Receptor occupancy and the stimulus were calculated according to classic receptor theory as described in Equations (15) and (16), respectively. In this example, K_{obs} was set at 3.39 × 10^{-7} M. D. Comparison between receptor occupancy and the stimulus function for a less efficacious agonist (ε = 0.21). Receptor occupancy and the stimulus were calculated according to classic receptor theory as described in Equations (15) and (16), respectively. In this example, K_{obs} was set at 7.76 × 10^{-6} M.

excess guanine nucleotide for two agonists of high (panel A) and low (panel B) intrinsic efficacy. The model was evaluated at a high concentration of guanine nucleotide, because GTP is present in the cell at concentrations that saturate G proteins. The solid curves show total receptor occupancy, and the dashed lines show the DR_s*GX component of agonist binding. It can be seen that when the concentration of GTP is high, agonist binding is consistent with a simple one-site model. Also, the concentration of agonist generating a half-maximal amount of the DR_s*GX component is equivalent to the observed dissociation constant (K_{obs}) of the agonist. For comparison, the relationship between receptor occupancy and the stimulus is shown for a highly efficacious agonist (ε = .69) and a less efficacious agonist (ε = .21) (Figures 2.16C and D, respectively). These latter curves were constructed using

Equations (15) and (16) to calculate receptor occupancy and the stimulus, respectively. It can be seen that the $DR_s{*}GX$ complex is a measure of the stimulus generated by the agonist, and that observed affinity and intrinsic efficacy at G protein-linked receptors obey the classic assumptions outlined in Equations (15) and (16).

This interpretation of the stimulus has been previously described in a study on the effects of GTP on agonist binding to M_2 muscarinic receptors and their inhibition of adenylyl cyclase activity in the rabbit myocardium.[36] Specifically, it was stated that the rate of activation of the G protein is proportional to the amount of agonist bound in the form of quaternary complex (*DRGX*). This statement has been misinterpreted by Waelbroeck[48] who assumed that the intent of the statement was to identify the *DRGX* complex as the active state of the G protein. On the contrary, our intent was to identify the *DRGX* complex as the precursor of the activated G protein, that is, the receptor complex from which free G_α-GTP and $\beta\gamma$ subunits are liberated. Our interpretation can readily account for the catalytic activation of G proteins by receptors in contrast to Waelbroek's assessment of our model.

2.4.2.1 Intrinsic Efficacy and the Ternary Complex Model

Since the ternary complex model is consistent with classic receptor theory, we can apply the model to the estimation of intrinsic efficacy and determine what parameters in the model correspond to intrinsic efficacy. First, we need to generate theoretical concentration-response curves using the ternary complex model and then apply Furchgott's method for estimation of relative efficacy.

The generation of a response by an agonist can be envisioned as a two-step process. First, the agonist generates a stimulus through the receptor, and then the transduction mechanisms in the cell convert the stimulus into a response. This second step is called the stimulus–response relationship. Given the complexity of the transduction mechanisms inside the cell, it might seem hopeless to try to represent them mathematically. However, there is an easy solution to this problem. The answer lies in the observation that most log concentration–response curves for agonists obey a logistic equation; that is, they resemble a sigmoid curve. This means that the entire train of events characteristic of the agonist's response (r) can be described by a simple equation expressing response as a function of the agonist concentration:

$$r = \frac{D^n \cdot E_{\text{max}}}{D^n + EC_{50}^n} \tag{17}$$

in which E_{max} denotes the maximal response, EC_{50} denotes the concentration of agonist eliciting a half-maximal response, and n is an exponent that governs the steepness of the concentration-response curve. If the entire train of events representing the agonist's response can be described by Equation (17) and the initial step in the process (i.e., the stimulus) can be described by Equation (16), then it can be shown that the events after the initial step (i.e., the stimulus–response relationship) must obey the following logistic equation:

$$r = \frac{S^m}{S^m + K_S^m} \tag{18}$$

in which S denotes the stimulus, K_S denotes the sensitivity of the transduction mechanism, and m is an exponent governing slope of the stimulus–response curve.[46,49-51] Thus, if we substitute the equation for the active state of the quaternary complex (i.e., Equation (A72), Appendix) for the stimulus in Equation (18), we can generate theoretical concentration response curves based on the ternary complex model.

Figure 2.17 shows the results of these calculations. The concentration–response curves for three agonists (*X*, *Y*, and *Z*) having the same affinity for the ground state of the receptor (K_a), but different affinities for the active state (aK_a) are shown in Figures 2.17A and B. Panel A shows the behavior of the agonists in a highly sensitive

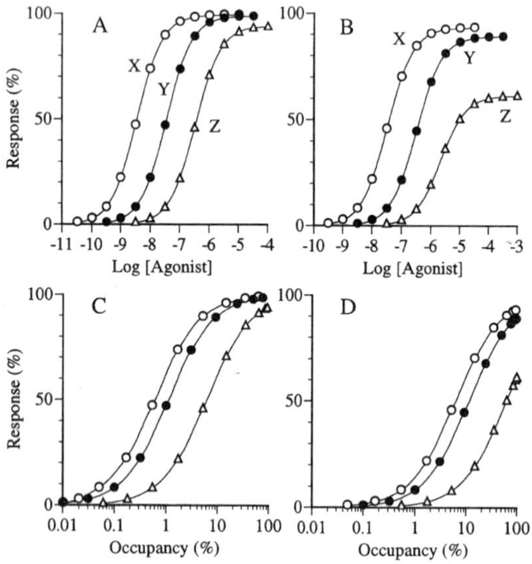

FIGURE 2.17 The effect of the ratio of microscopic affinity constants (i.e., aK_d/K_a or simply a) of an agonist for the ground (K_a) and active (aK_a) conformations of the receptor on the occupancy–response relationship. Theoretical concentration–response curves were generated for agonists X, Y, and Z using Equation (18) in the text, with Equation (A72) in the Appendix substituted for the value of the stimulus. The values of the microscopic constants of the agonists for the receptor conformations were the same except for the value of a, which was set at 10^6, 10^5, and 10^4 for agonists X, Y, and Z, respectively. A. Theoretical concentration–response curves for agonists X, Y, and Z in a highly sensitive system (i.e., $K_S = 0.001$ in Equation (18). B. Theoretical concentration–response curves for agonists X, Y, and Z in a less sensitive system where agonist Z behaves as a partial agonist (i.e., $K_S = 0.01$). C. Occupancy–response relationships of the agonists in the more sensitive system shown in A. D. Occupancy–response relationships of the agonists in the less sensitive system shown in B. The microscopic constants of the agonists for the receptor conformations were: $b = 10^5$, $c = 4 \times 10^{-4}$, $K_a = 1.6 \times 10^5$ M^{-1}, $K_b = 7 \times 10^{-3}$, $K_c = 8 \times 10^{-6}$, and $K_d = 10^8$ M^{-1}. The value of G_T/R_T was 0.8. The concentration of GTP was 10^{-3} M.

system (i.e., $K_S << 1$) where they all behave as full agonists. Panel B represents a less sensitive system (i.e., $K_S = 1.0$) where the less efficacious agonist exhibits a lower maximal response. Thus, in panel A the agonists exhibit a receptor reserve, whereas in panel B they do not. Figure 2.17C shows the responses in panel A plotted as a function of receptor occupancy for each agonist. An analogous plot is shown in panel D for the responses in panel B. It can be seen in panel C that the occupancy–response curves for agonists Y and Z are shifted to the right of that for X approximately 1.7- and 8.8-fold, respectively. Similar behavior is shown in panel D except that the occupancy–response relationship of the least efficacious agonist does not reach the same maximum as that of the more efficacious agonists. Nevertheless, with regard to equivalent responses, the occupancy–response curves of agonists Y and Z are shifted to the right of that for X 1.7- and 8.8-fold, respectively. This behavior shows that in both systems, it requires 1.7- and 8.8-fold greater receptor occupancy by Y and Z, respectively, to elicit a response equivalent to that of the more efficacious agonist *X*. In other words, the intrinsic efficacies of Y and Z, expressed relative to that of *X*, are 1/1.7 (0.58) and 1/8.8 (0.11), respectively. This method for estimating the relative efficacy of an agonist was first described by Furchgott.[47] We can represent this calculation mathematically by the following equation:

$$\varepsilon_A/\varepsilon_S = P_S/P_A \tag{19}$$

in which $\varepsilon_A/\varepsilon_S$ denotes intrinsic efficacy of A relative to that of S (i.e., relative efficacy of A), and P_S and P_A denote the receptor occupancies by S and A that yield equivalent responses.

The question then arises, as to what parameters in the ternary complex model determine intrinsic efficacy. The answer lies in how the theoretical concentration–response curves shown in Figures 2.17A and B were generated. These curves were generated using Equation (18), with the function for the active quaternary complex (DR_s*GX) substituted for the stimulus. Since intrinsic efficacy is proportional to the maximal stimulus, then the intrinsic efficacies of agonists X, Y, and Z should be proportional to the maximal value of $DR_s{*}GX$ ($DR_s{*}GX_{max}$). Accordingly, the $DR_s{*}GX_{max}$ values for agonists Y and Z, expressed relative to that of agonist X are 0.58 and 0.11, respectively. These values are identical to the relative efficacy values estimated for the agonists in the previous paragraph using the method of Furchgott. The question now is, what microscopic affinity constants in the ternary complex model determine DR_s*GX_{max}? Again, the answer lies in how the theoretical concentration–response curves were generated. These curves were constructed with the ratio of microscopic affinity constants for the ground and active states of the receptor (aK_d/K_a) set equal to 10^6, 10^5, and 10^4 for agonists X, Y, and Z, respectively. This ratio (*a*) determines the value of the α parameter in the ternary complex model. Accordingly, the values of α for agonists Y and Z expressed relative to that of X were 0.50 and 0.083. These values are similar to the relative efficacy values calculated above. Thus, it can be seen that intrinsic efficacy is nearly proportional to α in the simple ternary complex mode. This parameter is a measure of the selectivity of the agonist for the ternary complex (*DRG*) as compared to the *DR* complex. The parameter that measures the ability of the agonist to generate the ternary complex is αK_2:

$$D + R + G \overset{K_1}{\longleftrightarrow} DR + G \overset{\alpha K_2}{\longleftrightarrow} DRG \tag{20}$$

Since K_2 is a parameter of the system and is independent of the agonist, then intrinsic efficacy is also nearly proportional to αK_2.

Figure 2.18A shows the relationship between the ratio of microscopic affinity constants of an agonist for the ground and active conformations of the receptor (a) and the values of $DR_s{*}GX_{max}$ and αK_2. It can be seen that as the value of a increases from a low value, there is a proportional increase in both $DR_s{*}GX_{max}$ and αK_2. However, as the value of a increases further, $DR_s{*}GX_{max}$ and αK_2 reach a maximum. The plateau occurs because once the value of a exceeds a certain level, the receptor is essentially 100% activated, and a further increase in a will have no effect. At this point, $DR_s{*}GX_{max}$ is at its maximal value for the system. Similarly, αK_2 is at its maximal value because this parameter is a measure of the affinity of the DR complex for the G protein, and once the receptor is fully activated, it exhibits its highest affinity for the G protein. Therefore, a further increase in a will have no effect on αK_2 at this point. The plot in Figure 2.18A suggests that $DR_s{*}GX_{max}$ and αK_2 are proportional to one another. This nearly proportional relationship is illustrated in Figure 2.18B, which shows that the plot of αK_2 against $DR_s{*}GX_{max}$ yields a nearly straight line. Since $DR_s{*}GX_{max}$ is equivalent to observed intrinsic efficacy in the ternary complex model, Figure 2.18B shows that observed intrinsic efficacy is proportional to αK_2. In a prior study in which the αK_2 values of agonists were measured in binding assays on M_2 muscarinic receptors in the rabbit myocardium

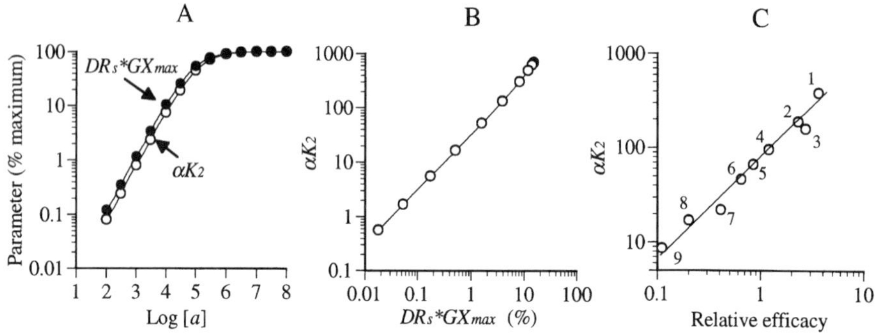

FIGURE 2.18 Relationships among relative efficacy, αK_2, and the ratio of microsopic affinity constants for the ground and active conformations of the receptor (a). The simulations in A and B were constructed using the same parameters used in Figure 2.17. A. The value of αK_2 and the maximal amount of active quaternary complex ($DR_s{*}GX$) is plotted against the ratio of microscopic affinity constants for the ground and active states of the receptor. B. αK_2 is plotted against $DR_s{*}GX$. C. The value of αK_2 for nine agonists was estimated in receptor binding assays on M_2 muscarinic receptors in the rabbit myocardium in the absence of GTP. Relative efficacy was estimated in adenylyl cyclase assays. The data are from Ehlert.[20] 1, oxotremorine-M; 2, carbachol; 3, *cis*-dioxolane; 4, oxotremorine; 5, (+) aceclidine; 6, (–) aceclidine; 7, N-methylaceclidine; 8, BM5; and 9, BOK1.

and compared with their relative efficacies for mediating an inhibition of adenylyl cyclase activity, exceptional agreement was noted (see Figure 2.18C).[20]

The relationships shown in Figure 2.18 have implications with regard to classic receptor theory and the definition of intrinsic efficacy. The mathematics of receptor theory provides us with an estimate of observed intrinsic efficacy. This value is proportional to both $DR_s{*}GX_{max}$ and αK_2 in the ternary complex model. Both of these parameters are dependent on the G protein; consequently, they are not strictly intrinsic properties of the agonist receptor complex. Therefore, these terms are a measure of *observed intrinsic efficacy* (ε_{obs}). When the $DR_s{*}GX_{max}$ and αK_2 values of an agonist are expressed relative to those of a standard agonist, they are nearly equivalent. These relative estimates are a measure of the *observed relative efficacy* of an agonist. To obtain an estimate of efficacy that is intrinsic to the agonist-receptor complex, we need to consider the third hierarchical level of modeling analysis (see Table 2.1). At the third level, intrinsic efficacy is most appropriately designated as a, the ratio of microscopic affinity constants of an agonist for the ground and active conformations of the receptor. This parameter is a true measure of the intrinsic efficacy of the agonist in the traditional sense. It is a measure of the agonist's ability to activate the receptor in isolation, and it is unaffected by the G protein. To avoid confusion, these two different meanings of intrinsic efficacy should be given different names. A simple solution is to rename Furchgott's definition of intrinsic efficacy as *observed intrinsic efficacy* (ε_{obs}) as described above, whereas the term *intrinsic efficacy* (ε) can be used to refer to a. This convention leads to the following relative terms when comparing different agonists; namely, *observed relative intrinsic efficacy* (*observed relative efficacy*) and *relative intrinsic efficacy* (*relative efficacy*). For the remainder of the chapter, this convention is used to designate the two types of intrinsic efficacy.

An interesting property of intrinsic efficacy noted above is that, in principle, this measure of the agonist's ability to "turn on" the receptor can exceed the capacity of the receptor to be turned on. For the specific example shown in Figure 2.18, once intrinsic efficacy (i.e., a) exceeds 10^6, a further increase is without effect on receptor activation because the receptor is already 100% activated. This relationship shows that it is theoretically possible for a group of the most efficacious agonists to exhibit the same maximal ceiling value of observed intrinsic efficacy. However, in the intact cell, it seems unlikely that high levels of receptor activation are achieved because of the large, negatively cooperative effects of guanine nucleotides on the receptor-G protein complex. For example, at M_2 muscarinic receptors, the most efficacious agonists exhibit different values for observed intrinsic efficacy when this parameter is measured using Furchgott's technique.[20] Also, the same group of efficacious agonists exhibit different αK_2 values when estimated through ligand binding analysis.[20] Finally, when the E_{max} values of the agonists are measured for an insensitive response, like phosphoinositide hydrolysis, the values are typically different, indicating differences in observed intrinsic efficacy.[52] All of these results suggest that the ceiling for observed relative efficacy has not been reached for the most efficacious agonists of muscarinic receptors. This situation implies that for most agonists, observed intrinsic efficacy and intrinsic efficacy are proportional as predicted in

Figure 2.18A for low values of a. In other words, both observed relative efficacy and relative efficacy are probably equal for most agonists in the intact cell with high concentrations of GTP.

2.4.2.2 Observed Affinity and the Ternary Complex Model

With regard to affinity, it is possible to use the technique of generating theoretical concentration-response curves to determine whether Furchgott's method of partial receptor inactivation is a valid method for estimating the K_{obs} of an agonist. This validation was reported by Black et al.[51,53] using the simple ternary complex as a measure of the stimulus. It is also possible to use the active quaternary complex as the equivalent of the stimulus and demonstrate that Furchgott's method provides an accurate measure of K_{obs} (analysis not shown). Finally, it has been shown that when Furchgott's method of partial receptor inactivation was used to estimate the observed affinities of agonists at M_2 muscarinic receptors mediating an inhibition of adenylyl cyclase, it was found that the estimates of K_{obs} were in excellent agreement with those measured by ligand binding in the presence of GTP.[54]

2.4.2.3 Observed Affinity and Intrinsic Efficacy are Interconvertible by GTP

As described above, traditional receptor theory considers affinity and intrinsic efficacy as intrinsic properties of the agonist-receptor complex in isolation, whereas the mathematics of receptor theory provides a measure of the observed affinity and relative efficacy of an agonist-receptor complex in its native context. One factor of the receptor's environment that can influence the observed affinity and intrinsic efficacy of the agonist-receptor complex is GTP.

The observed intrinsic efficacy of an agonist is proportional to the rate of guanine nucleotide exchange at maximal receptor occupancy. For a given agonist, this rate is proportional to the amount of the $DR_s{*}G$-GDP complex at saturating concentrations of agonist (see Figure 2.11). Ultimately, the amount $DR_s{*}G$-GDP complex is dependent on the amount of G-GDP complex, which in turn is dependent on the concentration of GTP. When the concentration of GTP is low, there is less G-GDP complex, and thus, less agonist-induced guanine nucleotide exchange at the G protein. Under these conditions, the agonist behaves as if it is less efficacious. However, with this decline in observed intrinsic efficacy, there is a corresponding increase in the observed affinity of the agonist. Observed affinity increases, because now more of the agonist-receptor-G protein complex is in the DRG complex as compared to the $DR_s{*}G$-GDP state. The agonist has lower affinity for the RG-GDP complex because GDP reduces the affinity of the agonist-receptor complex. This means that the microscopic affinity constant of the agonist for the RG-GDP complex is much smaller (lower affinity) than that of the RG complex. Nevertheless, the observed dissociation constant (K_{DRG}) for both states is the same at any given concentration of GTP due to the interconvertible nature of the states (see Figure 2.11, Table 2.8). The value of K_{DRG} increases (decrease in affinity) with an increase in the concentration of GTP. When there is no GTP, all of the agonist-receptor-G protein complexes are in the *DRG* complex, and there is essentially no stimulus. This complex has high affinity and stability, which is consistent with its inactivity. In order to generate a signal through a G protein-linked receptor, the G protein must shuttle back and forth between the receptor and its effector. This flow of information cannot happen when the G protein is bound tightly to the receptor. For example, it has been shown that fusion proteins constructed from the 5-HT receptor and $G_{\alpha i}$ can mediate an inhibition adenylyl cyclase by serotonin; however, the effect on cyclase depends on over-expression of the fusion protein.[55] When expressed at more physiological levels, fusion proteins of this sort have little effect on adenylyl cyclase.

We can investigate the effects of GTP on the agonist concentration–response curve using the technique applied in Figure 2.17. Accordingly, the stimulus is calculated at different concentrations of guanine nucleotide using equation A72 in the Appendix. This estimate of the stimulus is then substituted into Equation (18) to calculate the response. Figure 2.19 shows the results of these calculations for three different concentrations of GTP. Panel A shows that there is little effect of GTP on the agonist concentration–response curve. At the lowest concentration of GTP, both the maximal response and the EC_{50} value decrease so that there is little effect of GTP on the responses elicited by the agonist at low concentrations. Panel B shows the effects of GTP on receptor occupancy by the agonist. It can be seen that GTP causes a large 63-fold reduction in the observed affinity of the agonist as the concentration of GTP increases from 10^{-8} to 10^{-3} M. Panel C shows the occupancy-response relationships at the various concentrations of GTP. It can be seen that increasing the concentration of GTP shifts the occupancy–response relationship to the left and has the effect of increasing the observed intrinsic efficacy of the agonist.

From the foregoing, it is apparent that GTP has opposing effects on observed affinity and intrinsic efficacy. An increase in the concentration of GTP increases observed intrinsic efficacy and decreases observed affinity, whereas the converse is true for a decrease in GTP. The net effect is that the product of observed affinity and intrinsic efficacy remains constant over a range of concentrations of GTP. This property makes the concentration–response curve of the agonist insensitive to changes in the concentration of GTP when the transduction mechanism is highly sensitive (i.e., $K_S << 1$), like that shown in Figure 2.19. However, when the transduction mechanism is less sensitive (e.g., $K_S = 1$), then a decrease in the concentration of GTP causes a decrease in the maximal response and an increase in the potency of the agonist (i.e., decrease in EC_{50} value). Nevertheless, this effect of GTP still causes reciprocal changes in observed affinity and intrinsic efficacy such that their product remains constant over a range of concentrations of GTP. In a study of M_2 muscarinic receptor-mediated inhibition of adenylyl cyclase activity, it was shown that GTP had little effect on the concentration–response curves of highly efficacious agonists, but caused a huge decrease in the observed affinity of the agonists.[36] These results are consistent with the consequences of the ternary complex model as shown in Figure 2.19. Since neurotransmitters are usually highly efficacious and their signaling pathways are highly sensitive, the ternary complex model predicts that, under physiological conditions, signaling through G protein-linked receptors by endogenous neurotransmitters would be relatively insensitive to changes in the concentration of GTP in the cytosol.[36]

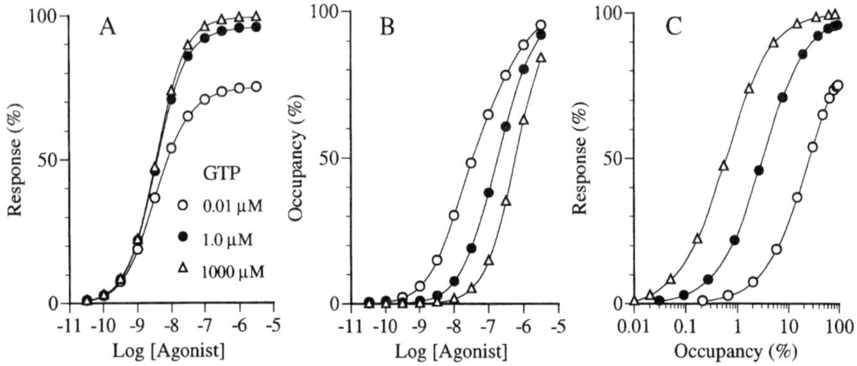

FIGURE 2.19 The influence of GTP on the concentration–response curve and receptor occupancy of a highly efficacious agonist. The effects of various concentrations of GTP on the concentration–response curve (A), the occupancy curve (B), and the occupancy–response relationship (C) of a highly efficacious agonist are shown. For these calculations, the microscopic constants of the agonist for the receptor conformations were the same as those used in Figure 2.10. Theoretical concentration–response curves were generated for the agonist using Equation (18) in the text, with Equation (A72) in the Appendix substituted for the value of the stimulus. The value of K_S in Equation (18) was 0.001.

2.4.2.4 The Product of Observed Affinity and Intrinsic Efficacy is a Useful Measure of Agonist Activity

As described above, each agonist-receptor complex has a unique capacity to express a certain combination of observed affinity and intrinsic efficacy. According to the consequences of the ternary complex model, these parameters are interchangeable so that a loss in one contributes to an increase in the other. The factor that governs the interchange is GTP. When there is no GTP, the agonist receptor complex exhibits its highest observed affinity, yet it is devoid of observed intrinsic efficacy. As the concentration of GTP increases, the agonist-receptor complex begins to display observed intrinsic efficacy as it accomplishes the work of guanine nucleotide exchange. However, it does so at the cost of a loss of observed affinity. The relationship between observed affinity and intrinsic efficacy is such that their product remains constant over most of the range in the concentration of GTP. From this viewpoint, each agonist-receptor complex exhibits a certain amount of agonistic activity and this activity can be defined as the product of the two interdependent parameters, observed affinity and intrinsic efficacy.

This method of defining agonist activity may seem trivial, because under physiological conditions, the cytosolic concentration of GTP is high, and therefore, observed affinity and intrinsic efficacy would be fixed at values unique to the agonist. Of what concern is the possibility of affinity-intrinsic efficacy interconversion when the concentration of GTP is probably always saturating inside the cell? There are two important issues to consider. First, the product of observed affinity and intrinsic efficacy is much easier to measure than each of the individual parameters (see discussion below), and second, many important biochemical measurements of agonist activity are made in broken cell preparations where the optimal conditions may require low concentrations of guanine nucleotide (e.g., agonist-stimulated [^{35}S]GTPγS binding). In the latter case, the individual estimates of observed affinity and intrinsic efficacy would be inaccurate estimates of their physiological values (i.e., those observed in the presence of high concentrations of GTP); nevertheless, their product would be accurate.

A simple method for estimating the product of observed affinity and intrinsic efficacy has been described. This estimate, termed the *IRA* value (intrinsic relative activity), represents the product of the observed affinity and intrinsic efficacy of an agonist (*A*) expressed relative to that of a standard agonist (*S*), and it can be calculated from the concentration–response curves of the two agonists.[52] In most instances, the *IRA* value is equivalent to the product of the ratios of maximal responses ($E_{max\text{-}A}/E_{max\text{-}S}$) and potencies ($EC_{50\text{-}S}/EC_{50\text{-}A}$):

$$IRA = \frac{E_{\max-A} \cdot EC_{50-S}}{E_{\max-S} \cdot EC_{50-A}} \tag{21}$$

In some situations, the calculation of *IRA* is more complicated, yet it still only requires analysis of the concentration–response curves. It has been shown that the *IRA* values of muscarinic agonists for stimulating contractions in the guinea pig ileum (M_3 response) are in excellent agreement with those measured in phosphoinositide assay on Chinese hamster ovary cells expressing the M_3 muscarinic receptor.[52] This agreement was observed even though most of the agonists behaved as highly potent full agonists in the guinea pig ileum and less potent partial agonists in the phosphoinositide assay.

2.5 ANALYSIS OF RECOMBINANT RECEPTORS

Perhaps the most unambiguous method for determining the pharmacological properties of agonist–receptor complexes is to examine their activity in cell lines transfected with the receptor gene of interest. This approach not only applies to the wild-type receptor, but also enables the investigation of various mutants and chimeras so that information about the structural determinants of agonist activity can be gained. In some instances it might be possible to analyze the agonist binding data in terms of the ternary complex model to estimate the observed intrinsic efficacy of the agonist-receptor complex (αK_2) as well as its observed affinity in the presence of GTP. However, the agonist binding properties of many G_q linked receptors do not exhibit two-site behavior or marked sensitivity to guanine nucleotides, making it impossible to estimate the parameters of the ternary complex model.[39-42] Moreover, there are ambiguities in the analysis of agonist binding at G_i- and G_o-linked receptors. The two-site nature of agonist binding at these receptors could be due to different pools of receptors having differential access to G proteins or to complexities in the G protein-receptor interaction (see prior discussion). This ambiguity could compromise the significance of the parameter estimates of the ternary complex model. Thus, a more robust approach is to measure binding properties in combination with a functional response such as GTPase activity, $[^{35}S]GTP\gamma S$ binding or second messenger levels (e.g., cAMP, inositol phosphates, and Ca^{2+}).

2.5.1 One Receptor, Many Agonists

The simplest way to estimate the pharmacological activity of agonists is to measure their concentration–response curves in a cell line transfected with the receptor of interest. This information is sufficient to estimate the *IRA* values of agonists. As described above, the *IRA* value is equivalent to the product of observed affinity and intrinsic efficacy expressed relative to that of another agonist. If the concentration response curve exhibits a Hill coefficient similar to one, then the measurement of *IRA* is extremely simple; it only requires estimates of EC_{50} and E_{max}. Even if the Hill slopes differ from one, it is still possible to estimate the *IRA* value, although regression analysis is required in this circumstance. If affinity and intrinsic efficacy are determined mainly by the receptor, then the *IRA* values of agonists should be relatively constant as long as the same receptor mediates the response. However, if G proteins modify pharmacological activity, then one might expect to observe different *IRA* values for the same agonist-receptor complex, depending upon the G protein with which the receptor couples.

Although the *IRA* value is extremely valuable and simple to estimate, it does not provide separate estimates of observed affinity and intrinsic efficacy, but rather their product. In order to obtain separate estimates, it is necessary to measure the observed dissociation constant of the agonist. This measurement could be made in competitive ligand binding experiments on homogenates of the transfected cells. The binding assays should be carried out in a buffer system similar to that of the functional assay. If the functional assay is measured in intact cells, then the binding assay should include excess GTP to mimic the conditions of the cytosol. Since most G protein-linked receptors exhibit simple one-site agonist binding curves in the presence of GTP, it should be relatively straightforward to estimate the K_{obs} from the binding experiment. Knowing the K_{obs}, it is possible to factor out the affinity component from the *IRA* value to obtain an estimate of the observed relative efficacy of an agonist ($\varepsilon_{obs\text{-}A}/\varepsilon_{obs\text{-}S}$):

$$\frac{\varepsilon_{obs-A}}{\varepsilon_{obs-S}} = IRA\frac{K_{obs-A}}{K_{obs-S}} \tag{22}$$

in which ($\varepsilon_{obs\text{-}A}/\varepsilon_{obs\text{-}S}$) denotes the observed intrinsic efficacy of *A* expressed relative to the standard agonist *S*. If the concentration–response curves have Hill slopes not much different from one, then Equation (21) can be substituted into Equation (22) yielding:

$$\frac{\varepsilon_{obs-A}}{\varepsilon_{obs-S}} = \frac{E_{\max-A}\cdot EC_{50-S}\cdot K_{obs-A}}{E_{\max-S}\cdot EC_{50-A}\cdot K_{obs-S}} \tag{23}$$

This method of estimating relative efficacy has been previously discussed by Ehlert et al.[52] and Trzeciakowski.[56] Alternatively, if the Hill slopes differ significantly from one, then the method for calculating the corrected *IRA* value described by Ehlert et al.[52] can be used in Equation (22) to estimate the relative efficacy of the agonist.

Alternatively, separate estimates of observed affinity and intrinsic efficacy can be obtained by the method of partial receptor inactivation.[46] This technique requires the measurement of concentration–response curves to the agonist before and after partial inactivation of the receptors with an irreversible antagonist. Analysis of the data yields an estimate of the observed affinity of the agonist. This estimate then enables the determination of the occupancy–response relationship (e.g., see Figures 2.17C and D) from which observed relative efficacy can be calculated. This method was originally invented by Furchgott,[47] and some important modifications of the method have been described, perhaps the most important of which is that described by Black and coworkers.[51,53] Their technique enables the estimation of observed affinity and relative efficacy by direct nonlinear regression analysis of the concentration–response curves. When expressed relative to other agonists, the τ parameter of their model value is equivalent to observed relative efficacy.

2.5.2 Many Receptors, One Agonist

Recombinant DNA technology has facilitated the exploration of the structural basis of receptor function. By measuring the concentration–response curve to a highly efficacious agonist in cells transfected with different receptor constructs, it should be possible to identify domains on the receptor that contribute to affinity and intrinsic efficacy. However, there are two potential caveats in these types of experiments. The first is that receptor expression may vary for different mutants, causing the sensitivity of the response to change. If the analysis fails to account for altered receptor expression, then estimates of observed relative efficacy can be erroneous. Second, it is essential to discriminate between the traditional concept of affinity and intrinsic efficacy and their observed values in these types of experiments.

This latter issue has been discussed by Colquhoun[57] in connection with ligand-gated ion channels. It is informative to illuminate this issue at G protein-linked receptors through the use of examples. Imagine a point mutation that reduces the affinity of an agonist for the active state of a G_i linked receptor. According to traditional receptor theory, agonist affinity would be unchanged, whereas intrinsic efficacy should be reduced. When measured in the presence of GTP, the observed affinity of the agonist would probably be unaffected or slightly reduced, leading to the conclusion that the decreased responsiveness of the receptor construct is due to diminished intrinsic efficacy. However, if binding affinity was measured in the absence of GTP, a decrease in observed affinity would be observed in the mutant. As described above, reducing the concentration of GTP causes a decrease in observed intrinsic efficacy and a corresponding increase in observed affinity. In other words, intrinsic efficacy is converted into affinity by the removal of GTP. Therefore a decrease in intrinsic efficacy is manifested as reduced affinity when measured in the absence of GTP. For this reason, it is essential to define observed affinity and intrinsic efficacy with respect to the concentration of GTP. The most appropriate concentration is an excess of GTP since this condition resembles the physiological situation in the intact cell.

A point mutation that has no influence on the structure of the ground and active conformations, but inhibits the agonist-induced conformational change, also has interesting consequences. This situation would yield results analogous to those just described. According to traditional receptor theory, the mutant has reduced intrinsic efficacy, but no change in affinity. The estimates of observed affinity and relative efficacy would be generally consistent with the traditional view as long as observed affinity is defined as that measured in the presence of excess GTP.

Finally, imagine a point mutation in the i3 loop that has no effect on the structure of the ground and active states of the receptor, nor on their ability to undergo a conformational change. Nevertheless, the point mutation hinders G protein coupling. According to a strict interpretation of traditional receptor theory, the point mutation should have no effect on affinity and intrinsic efficacy. It is only our ability to observe intrinsic efficacy through G protein coupling that has been perturbed. This point mutation would cause a decrease in the observed affinity of agonists in the absence of GTP, but little or no effect in the presence of GTP. However, the diminished G protein coupling of the agonist–receptor complex would be manifest as a decrease in observed intrinsic efficacy. Thus, the observed intrinsic efficacy is both a measure of the agonist's ability to activate the receptor as well as the ability of the active receptor conformation to interact with G proteins.

Since observed affinity and intrinsic efficacy are properties of the drug–receptor complex, they are dependent on both the agonist and the receptor. Consequently, changes in receptor structure may have differential effects on observed affinity and intrinsic efficacy, depending upon the agonist. After all, a mutant receptor is distinct from the wild type, and therefore, it should display a unique pharmacological profile, assuming that critical domains are modified in the mutant. Accordingly, it would be improbable that the wild type and mutant receptor would exhibit exactly the same rank order for observed affinity and relative efficacy for all agonists. Since the picture that emerges from site-directed mutagenesis studies depends on the agonist, the natural neurotransmitter for the receptor is perhaps the most important agonist to investigate. All data with the neurotransmitter have relevance to the physiological function of the receptor, whereas data with a particular agonist might not be relevant to the neurotransmitter or to other agonists.

To measure the observed affinity and intrinsic efficacy of an agonist for different receptor mutants, methods analogous to those described in the preceding section can be used provided that corrections for differences in receptor expression are made. For each mutant, the relative efficacy of an agonist can be compared with the wild type receptor. Thus, in this situation, the intrinsic efficacy of the receptor is expressed relative to the wild type, keeping the agonist constant.

The simplest approach is to measure the product of observed affinity and intrinsic efficacy of the mutant relative to the wild type for the same agonist. This estimate (*RAM*: relative activity of the mutant) can be calculated from the agonist concentration–response curves in the mutant and wild type receptor in a manner analogous to the *IRA* calculation with a correction factor for differences in receptor density:

$$RAM = \frac{E_{\max-M} \cdot EC_{50-WT} \cdot B_{\max-WT}}{E_{\max-WT} \cdot EC_{50-M} \cdot B_{\max-M}} \tag{24}$$

In this equation $EC_{50\text{-}WT}$ and $EC_{50\text{-}M}$ denote the EC_{50} values of the agonist for eliciting a response in cells transfected with the wild type and mutant receptors, and $E_{\max\text{-}WT}$ and $E_{\max\text{-}M}$ denote the corresponding maximal responses. The receptor densities of the wild type and mutant receptors are denoted by $B_{\max\text{-}WT}$ and $B_{\max\text{-}M}$, respectively. Because there are potential differences in the cellular localization of the mutant and wild type receptors, it is important to measure only functional receptors expressed on the cell surface. This could be achieved by measuring the receptor density in intact cells with a membrane-impermeable ligand. The use of Equation (24) is subject to the same limitations as the *IRA* calculation [Equation (21)]. The concentration–response curve of the agonists should have Hill coefficients similar to one, and the mutant receptor must signal through the same transduction mechanism as the wild type receptor. In situations where the concentration–response curves of the agonist in the cells expressing the wild type and mutant receptor exhibit dissimilar Hill coefficients, the *RAM* value can be estimated in a manner analogous to that described for the *IRA* value under this condition, with the addition of the correction factor for differences in receptor density:

$$RAM = IRA \cdot \frac{B_{\max-WT}}{B_{\max-M}} \tag{25}$$

To obtain separate estimates of the observed affinity and intrinsic efficacy of the agonist at the mutant receptor relative to the wild type, it is necessary to measure the observed dissociation constant of the agonist at the two receptors. This can be done as described in the preceding sections. Knowing observed affinity, it can be factored out of the *RAM* value in a manner similar to that described for the *IRA* calculation to obtain an estimate of the observed intrinsic efficacy of the agonist at the mutant relative to the wild type ($\varepsilon_{obs\text{-}M}/\varepsilon_{obs\text{-}WT}$):

$$\frac{\varepsilon_{obs-M}}{\varepsilon_{obs-WT}} - RAM \frac{K_{obs-M}}{K_{obs-WT}} \tag{26}$$

In situations where the Hill coefficients of the agonist concentration–response curves are not significantly different from one, Equation (24) can be substituted into Equation (26) to yield:

$$\frac{\varepsilon_{obs-M}}{\varepsilon_{obs-WT}} = \frac{E_{\max-M} \cdot EC_{50-WT} \cdot B_{\max-WT} \cdot K_{obs-M}}{E_{\max-WT} \cdot EC_{50-M} \cdot B_{\max-M} K_{obs-WT}} \tag{27}$$

Caution should be exercised in applying the equations for estimation of *RAM*. One can never be certain that the estimate of $B_{\max\text{-}M}/B_{\max\text{-}WT}$ is a reliable measure of the relative amount of functional receptors in transfection experiments, even if care is taken to measure only cell surface receptors. There could be differences in the compartmentalization of receptors on the cell surface that influence G protein coupling. For example, if 100% of the wild type receptors on the cell surface are functionally coupled, whereas only 50% of the cell-surface mutant receptors are coupled, then the estimate of $B_{max\text{-}M}/B_{max\text{-}WT}$ will be twofold greater than the true value. There is no simple way to correct for this potential problem.

Nevertheless, it should be possible to test whether the measured binding capacities of a particular receptor construct are valid estimates in different cell lines derived from the same parent cell line by transfecting the same receptor at various densities. According to theory, the *IRA* value of a standard agonist in one cell line expressed relative to itself in another cell line expressing the same receptor should be equivalent to the ratio of B_{max} values of the two cell lines:

$$\frac{B_{max}{}^*}{B_{max-control}{}^*} = IRA_{control} \tag{28}$$

In this equation, $B_{max}{}^*$ and $B_{max\text{-}control}{}^*$ denote the theoretical binding capacities of the receptor in two different cell lines with one arbitrarily defined as the control, and $IRA_{control}$ denotes the *IRA* value of an agonist in one cell line expressed relative to itself in the control cell line. If the concentration–response curves of the agonist exhibit Hill coefficients similar to one in both cell lines, then Equation (28) simplifies to:

$$\frac{B_{max}{}^*}{B_{max-control}{}^*} = \frac{E_{max} \cdot EC_{50-control}}{E_{max-control} \cdot EC_{50}} \tag{29}$$

in which E_{max} and EC_{50} denote the maximal response and EC_{50} value of the agonist in one cell line, and $E_{max\text{-}control}$ and $EC_{50\text{-}control}$ denote the corresponding values in the control cell line. Thus, one can use Equation (29) to determine if the theoretical ratio of functional receptor densities ($B_{max}{}^*/B_{max\text{-}control}{}^*$) is equivalent to the actual ratio of receptor densities measured in a ligand binding assay ($B_{max}/B_{max\text{-}control}$). To illustrate the application of this equation, it is useful to consider an example from the literature in which the EC_{50} value and maximal response of carbachol for eliciting phosphoinositide hydrolysis were measured in seven B82 fibroblast cell lines expressing different levels of M_1 muscarinic receptors.[58] In this study, it was noted that the maximal response and potency ($1/EC_{50}$) of carbachol increased with an increase in the expression of receptors on the cell surface. The binding capacity of [^{3}H]N-methylquinuclidinyl benzilate ([^{3}H]NMQNB) increased 22-fold across the cell lines with a corresponding 8.6-fold increase in E_{max} (maximal response minus basal) and a 7.4-fold variation in potency. Since the concentration–response curves for carbachol-stimulated phosphoinositide hydrolysis exhibited Hill coefficients of approximately one, it is appropriate to use Equation (29) to estimate the theoretical ratio of functional binding capacities in the cell lines. Using the LK3-1 cell line as the control, the following ratios of 1.0, 0.78, 1.1, 4.3, 5.2, 28, and 29 can be calculated for $B_{max}{}^*/B_{max\text{-}control}{}^*$ in the LK3-1, LK3-4, LK7-6, LK3-7, LK7-2, LK3-3, and LK3-8, respectively. The actual binding capacities of [^{3}H]NMQNB, expressed relative to the LK3-1 cell line, were 1.0, 1.5, 2.5, 8.0, 12, 20, and 22. It can be seen that there is general agreement (within a factor of two) between the two different estimates suggesting that the binding capacity of [^{3}H]NMQNB to intact cells represents a reliable measure of the density of functional receptors on the different cell lines.

Another useful method for estimation of the pharmacological properties of agonists in cell lines expressing recombinant receptors has been described by Lu and Hulme.[59] The method is based on the simple ternary complex model, and has been applied to G_q-linked M_1 muscarinic receptors. The method is based on the assumption that the response is proportional to the amount of receptor in the G protein-coupled complex, and that the observed binding affinity of agonists is a measure of K_1. This method yields estimates of relative efficacy that are the same as those described previously [Equations (22) and (27)], although the latter methods are not based on the assumptions of the method of Lu and Hulme.[59] Regardless, these assumptions do not appear to limit the usefulness of Lu and Hulme's method, which indicates the inherent robustness of an analysis based on the ternary complex model.

2.6 FINAL REMARKS

This chapter was written with the conviction that the ternary complex model and classic receptor theory are largely compatible and have widespread application in the analysis of G protein-linked receptors. Undoubtedly, we will be modifying these models for years to come; nevertheless, the essence of these approaches explains many phenomena of G protein-linked receptors.

Much confusion exists in the literature because of a lack of clear distinction among the different levels of modeling analysis (see Table 2.1). Often, observed parameters (level 1 parameters) are confused with microscopic parameters (level 2 and 3 parameters), and the recognition of level 3 parameters has often been interpreted as evidence that a level 2 analysis is flawed (e.g., the simple ternary complex model is invalid because it does not identify ground and active conformations of the receptor). As shown throughout this chapter, the three hierarchical levels of analysis are internally consistent. Finally, sometimes variables are used inconsistently within receptor models. For example, a level 3 designation of a receptor state (e.g., active state, R*) is often used as such to designate the free receptor; however, when the receptor is bound with ligand and G protein, the same state is sometimes shown to exhibit different affinities. This practice shows that the designated state is not really a single state at all, but rather an equilibrium between two or more states. Thus, in this situation, the purported level 3 analysis is actually a level 2 analysis.

Undoubtedly, there are those who will assert that a two-state analysis is invalid because there is evidence for additional states of the receptor. My intent in using a two-state analysis in this chapter is not to proclaim that there are only two states of the receptor, but rather to illustrate how the ternary complex model and classic receptor theory can be reconciled with a model incorporating more than one receptor conformation. The same analysis could have been done with more than two states. Such a model would not invalidate the internal consistency of the hierarchical analysis. Kenakin[60] and Leff et al.[61] have provided excellent descriptions of models incorporating different activated receptor conformations with differential selectivity for G proteins.

An important topic not addressed in this chapter is receptor oligomerization. There is now convincing evidence that some G protein-linked receptors form homo- and hetero-oligomers. A striking example are κ and δ opiate receptors, which form at least a heterodimer that has unique pharmacological properties and mediates an inhibition of adenylyl cyclase activity.[62] The heterodimer appears to exhibit positive cooperativity in the binding of selective agonists to each monomer. Presumably, the oligomeric receptor interacts with the G protein to trigger a stimulus. The simple ternary complex model does not show oligomerization at the level of the receptor or G protein. Nevertheless, we should not hasten to discard the model. The essence of the model is that it shows how the agonist–receptor complex associates with a G protein and causes an increase in the dissociation rate of GDP. This is how the receptor works. We may have to refine the model to show that the receptor complex is actually an oligomer, and perhaps, more than one G protein hetero-oligomer may interact with the oligomeric receptor complex.

This latter point raises the issue of how we should judge the validity of receptor models. Recently, much has been written about the flaws in classic receptor theory. This position is based on the assumption that the authors of the theory intended to measure microscopic constants. Since classic theory fails at this measurement, it is considered by some to be flawed. This line of reasoning tells us what classic receptor theory does not do. There is, by the way, an infinitely long list of other things the theory does not do. It is difficult to imagine how a method for estimating microscopic constants at G protein-linked receptors could have been derived at a time when G proteins had yet to be discovered. Moreover, it is hard to imagine that Furchgott had intended to measure microscopic constants because he has previously distinguished between his measure of the agonist dissociation constant (K_A) and the microscopic constants of the ground and active conformations of the receptor.[63] The ingenious approach used by del Castillo and Katz[64] to measure microscopic constants at nicotinic receptors, for example, is inappropriate for estimation of microscopic constants at G protein-linked receptors because it does not consider the second hierarchical level of analysis. In my view, there are two issues that should be distinguished when assessing the validity of receptor models — the intent of the author and the usefulness of the model. Although the former issue often has great sociological impact, it has virtually no bearing on the latter issue. To consider the latter issue, we can assess a model in the same way we determine the function of genes and proteins in biology. After all, a model in written form is a product of the human mind; and thus, it is a complex *extended phenotype* of the human genome (see Dawkins[65] for the definition of extended phenotype). When we assess the function of a gene or protein, the concept of intent or authorship has no place. We simply want to know how the protein works. Thus, in the same light, we should consider the function of classic receptor theory, or rather, how the theory can be applied. It is my hope that this chapter illustrates the usefulness of classic receptor theory in the measurement of observed pharmacological parameters, and how these parameters are related to more fundamental microscopic constants.

2.7 APPENDIX

In this analysis, observed dissociation constants are defined in units of molarity, whereas the microscopic affinity constants for complexes and states of the receptor are defined in units of inverse molarity.

2.7.1 Ternary Complex Model with Guanine Nucleotide

The equations below are derived to describe the binding of a drug (D) to a receptor (R) that can associate with a G protein (G) that can bind guanine nucleotide (X). The microscopic affinity constants of D for the various receptor complexes (R, RG, and RGX) are:

$$K_1 = [DR]/[D][R] \tag{A1}$$

$$\alpha K_1 = [DRG]/[D][RG] \tag{A2}$$

$$\alpha\gamma K_1 = [DRGX]/[D][RGX] \tag{A3}$$

The microscopic affinity constants of guanine nucleotide (X) for the various complexes of G protein are given by:

$$K_3 = [GX]/[X][G] \tag{A4}$$

$$\beta K_3 = [RGX]/[X][RG] \tag{A5}$$

$$\beta\gamma K_3 = [DRGX]/[X][DRG] \tag{A6}$$

Because the interaction between R and G occurs within the membrane, their local concentrations are independent of their total concentration in solution. Consequently, it is useful to normalize their affinity constants relative to the total receptor concentrations ($[R_T]$):

$$K_2 = [RG][R_T]/[R][G] \tag{A7}$$

$$\alpha K_2 = [DRG][R_T]/[DR][G] \tag{A8}$$

$$\beta K_2 = [RGX][R_T]/[R][GX] \tag{A9}$$

$$\alpha\beta\gamma K_2 = [DRGX][R_T]/[DR][GX] \tag{A10}$$

The following conservation of mass equations apply:

$$[R_T] = [R] + [DR] + [DRG] + [DRGX] + [RG] + [RGX] \tag{A11}$$

$$[G_T] = [G] + [GX] + [RGX] + [DRGX] + [RG] + [DRG] \tag{A12}$$

The ratio of G_T/R_I is given by:

$$r = [G_T]/[R_T] \tag{A13}$$

2.7.1.1 Binding of Drug to Receptor

The equation describing the binding of D to the receptor is given by:

$$\frac{[DR]+[DRG]+[DRGX]}{[R_T]}=\frac{[D]}{[D]+[K_{obs}]} \tag{A14}$$

In this equation, K_{obs} denotes the observed dissociation constant of the drug. It is defined in terms of the microscopic affinity constants:

$$\begin{aligned}
K_{obs} &= \frac{1+G_F K_2(1+[X]\beta K_3)}{K_1+G_F\alpha K_1 K_2(1+[X]\beta\gamma K_3)} \\
G_F &= \frac{B+\sqrt{B^2+4AC}}{2A} \\
A &= 1+[X]K_3 \\
B &= r-1-AK_G \\
C &= rK_G \\
K_G &= \frac{1+[D]K_1}{K_2+[D]\alpha K_1 K_2+[X]\beta K_2 K_3+[D][X]\alpha\beta\gamma K_1 K_2 K_3}
\end{aligned} \tag{A15}$$

The equation describing the amount of D bound in the form of the quaternary complex is:

$$\frac{[DRGX]}{[R_T]}=\frac{[D]}{[D]+K_{DRGX}} \tag{A16}$$

In this equation, K_{DRGX} denotes the observed dissociation constant of the *DRGX* complex. It is defined in terms of the microscopic affinity constants:

$$K_{DRGX}=\frac{1+[D]K_1+G_F K_2+[D]G_F\alpha K_1 K_2+[X]G_F\beta K_2 K_3}{[X]G_F\alpha\beta\gamma K_1 K_2 K_3} \tag{A17}$$

In this equation, G_F is defined as in Equation (A15). The equation describing the amount of D bound in the form of ternary complex is:

$$\frac{[DRG]}{[R_T]}=\frac{[D]}{[D]+K_{DRG}} \tag{A18}$$

In this equation, K_{DRG} denotes the observed dissociation constant of the *DRG* complex. It is defined in terms of the microscopic affinity constants:

$$K_{DRG} = \frac{1+[D]K_1 + G_F K_2 + [X]G_F \beta K_2 K_3 + [D][X]G_F \alpha\beta\gamma K_1 K_2 K_3}{G_F \alpha K_1 K_2} \quad \text{(A19)}$$

In this equation, G_F is defined as in Equation (A15). The equation describing the amount of *D* bound in the form of the *DR* complex is:

$$\frac{[DR]}{[R_T]} = \frac{[D]}{[D]+K_{DR}} \quad \text{(A20)}$$

In this equation, K_{DR} denotes the observed dissociation constant of the *DR* complex. It is defined in terms of the microscopic affinity constants:

$$K_{DR} = \frac{1+G_F K_2 + [D]G_F \alpha K_1 K_2 + [X]G_F \beta K_2 K_3 + [D][X]G_F \alpha\beta\gamma K_1 K_2 K_3}{K_1} \quad \text{(A21)}$$

In this equation, G_F is defined as in Equation (A15).

2.7.1.2 Binding of Guanine Nucleotide to G Protein

The equation describing the binding of *X* to the G protein is given by:

$$\frac{[GX]+[RGX]+[DRGX]}{[G_T]} = \frac{[X]}{[X]+K_X} \quad \text{(A22)}$$

In this equation, K_X denotes the observed dissociation constant of the guanine nucleotide. It is defined in terms of the microscopic affinity constants:

$$K_X = \frac{1+R_F K_2(1+[D]\alpha K_1)}{K_3 + R_F \beta K_2 K_3(\alpha + [D]\alpha\gamma K_1)}$$

$$R_F = \frac{B+\sqrt{B^2 + 4AK_R}}{2A}$$

$$A = 1+[D]K_1 \quad \text{(A23)}$$

$$B = 1-r-K_R(1+[D]K_1)$$

$$K_R = \frac{1+[X]K_3}{K_2 + [X]\beta K_2 K_3 + [D]\alpha K_1 K_2 + [D][X]\alpha\beta\gamma K_1 K_2 K_3}$$

2.7.2 Conformational Analysis of the Ternary Complex Model

Each receptor complex described in the previous section can be divided into two components representing the ground (R_s) and active (R_s*) states of the receptor. The microscopic affinity constants of the two states of the various receptor complexes in the ternary complex model are defined below. Also described are the microscopic constants describing the equilibrium between the two states of the receptor complexes:

$$K_a = [DR_s]/[D][R_s] \tag{A24}$$

$$K_a = [DR_sG]/[D][R_sG] \tag{A25}$$

$$K_a = [DR_sGX]/[D][R_sGX] \tag{A26}$$

$$aK_a = [DR_s*]/[D][R_s*] \tag{A27}$$

$$aK_a = [DR_s*G]/[D][R_s*G] \tag{A28}$$

$$aK_a = [DR_s*GX]/[D][R_s*GX] \tag{A29}$$

$$K_b = [R_T][R_sG]/[R_s][G] \tag{A30}$$

$$K_b = [R_T][DR_sG]/[DR_s][G] \tag{A31}$$

$$K_b = [R_T][R_sGX]/[R_s][GX] \tag{A32}$$

$$K_b = [R_T][DR_sGX]/[DR_s][GX] \tag{A33}$$

$$bK_b = [R_T][R_s*G]/[R_s*][G] \tag{A34}$$

$$bK_b = [R_T][DR_s*G]/[DR_s*][G] \tag{A35}$$

$$bcK_b = [R_T][R_s*GX]/[R_s*][GX] \tag{A36}$$

$$bcK_b = [R_T][DR_s*GX]/[DR_s*][GX] \tag{A37}$$

$$K_c = [R_s*]/[R_s] \tag{A38}$$

$$aK_c = [DR_s*]/[DR_s] \tag{A39}$$

$$bK_c = [R_s*G]/[R_sG] \tag{A40}$$

$$abK_c = [DR_s*G]/[DR_sG] \tag{A41}$$

$$bcK_c = [R_s{*}G]/[R_sG] \tag{A42}$$

$$abcK_c = [DR_s{*}GX]/[DR_sGX] \tag{A43}$$

$$K_d = [GX]/[X][G] \tag{A44}$$

$$K_d = [R_sGX]/[X][R_sG] \tag{A45}$$

$$K_d = [DR_sGX]/[X][DR_sG] \tag{A46}$$

$$cK_d = [R_s{*}GX]/[R_s{*}G][X] \tag{A47}$$

$$cK_d = [DR_s{*}GX]/[DR_s{*}G][X] \tag{A48}$$

It is possible to define the microscopic affinity constants for the different receptor complexes in the ternary complex model in terms of the microscopic constants for the different states of the receptor just described:

$$K_1 = \frac{K_a + aK_aK_c}{K_c + 1} \tag{A49}$$

$$\alpha K_1 = \frac{K_a + abK_aK_c}{bK_c + 1} \tag{A50}$$

$$\alpha\gamma K_1 = \frac{K_a + abcK_aK_c}{bcK_c + 1} \tag{A51}$$

$$K_2 = \frac{K_b + bK_bK_c}{K_c + 1} \tag{A52}$$

$$\alpha K_2 = \frac{K_b + abK_bK_c}{aK_c + 1} \tag{A53}$$

$$\beta K_2 = \frac{K_b + bcK_bK_c}{K_c + 1} \tag{A54}$$

$$\alpha\beta\gamma K_2 = \frac{K_b + abcK_bK_c}{aK_c + 1} \tag{A55}$$

$$K_3 = K_d \tag{A56}$$

$$\beta K_3 = \frac{K_d + bcK_cK_d}{bK_c + 1} \tag{A57}$$

$$\beta\gamma K_3 = \frac{K_d + abcK_cK_d}{abK_c + 1} \tag{A58}$$

The estimates of the cooperativity values (i.e., α, β, and γ) in the ternary complex model can also be defined in terms of the microscopic constants of the receptor states:

$$\alpha = \frac{(K_a + abK_aK_c)(K_c + 1)}{(K_a + aK_aK_c)(bK_c + 1)} \tag{A59}$$

$$\beta = \frac{1 + bcK_c}{1 + bK_c} \tag{A60}$$

$$\gamma = \frac{(K_a + abcK_aK_c)(bK_c + 1)}{(bcK_c + 1)(K_a + abK_aK_c)} \tag{A61}$$

The equation describing the binding of D to the receptor in the absence of GTP is:

$$\frac{[DR_s] + [DR_s{}^*] + [DR_sG] + [DR_s{}^*G]}{[R_T]} = \frac{[D]}{[D] + [K_{obs}]} \tag{A62}$$

In this equation, K_{obs} denotes the observed dissociation constant of the drug. In the absence of guanine nucleotide, the equation for K_{obs}, expressed in terms of the microscopic constants of the states of the receptor, is:

$$\begin{aligned}
K_{obs} &= \frac{1 + K_c + G_FK_b + G_FbK_bK_c}{K_a + aK_aK_c + G_FK_aK_b + G_FabK_aK_bK_c} \\
G_F &= \frac{B + \sqrt{B^2 + 4C}}{2} \\
B &= r - 1 - K_G \\
C &= rK_G \\
K_G &= \frac{1 + Kc[D]Ka + [D]aKaKc}{K_b + [D]K_aK_b + bK_bK_c + [D]abK_aK_bK_c}
\end{aligned} \tag{A63}$$

The equation describing the amount of drug bound to the active state in the form of ternary complex is:

$$\frac{[DR_s * G]}{[R_T]} = \frac{[D]}{[D] + K_{DR_s * G}} \tag{A64}$$

In this equation, K_{DR_s*G} denotes the observed dissociation constant of the DR_s*G complex. It is defined in terms of the microscopic affinity constants:

$$K_{DR_s*G} = \frac{1 + K_c + [D]K_a + G_F K_b + [D]aK_a K_b + G_F[D]K_a K_b + bK_b K_c}{G_F bK_a K_b K_c} \tag{A65}$$

In this equation, G_F is defined as in Equation (A63). The equation describing the amount of D bound to the active state of the receptor is:

$$\frac{[DR_s *]}{[R_T]} = \frac{[D]}{[D] + K_{DR_s *}} \tag{A66}$$

In this equation, K_{DR_s*} denotes the observed dissociation constant of the DR_s* complex. It is defined in terms of the microscopic affinity constants:

$$K_{DR_s*} = \frac{1 + K_c + [D]K_a + G_F K_b + [D]G_F K_a K_b + G_F bK_b K_c + [D]G_F abK_a K_b K_c}{aK_a K_c} \tag{A67}$$

In this equation, G_F is defined as in Equation (A63). The equation describing the amount of D bound to the ground state of the receptor in the form of ternary complex is:

$$\frac{[DR_s G]}{[R_T]} = \frac{[D]}{[D] + K_{DR_s G}} \tag{A68}$$

In this equation, K_{DR_sG} denotes the observed dissociation constant of the DR_sG complex. It is defined in terms of the microscopic affinity constants:

$$K_{DR_sG} = \frac{1 + K_c + [D]K_a + [D]aK_a K_c + G_F K_b + G_F bK_b K_c + G_F[D]abK_a K_b K_c}{G_F K_a K_b} \tag{A69}$$

In this equation, G_F is defined as in Equation (A63). The equation describing the amount of D bound to the ground state of the receptor is:

$$\frac{[DR_s]}{[R_T]} = \frac{[D]}{[D] + K_{DR_s}} \tag{A70}$$

In this equation, K_{DR_sG} denotes the observed dissociation constant of the DR_s complex. It is defined in terms of the microscopic affinity constants:

$$K_{DR_s} = \frac{1 + K_c + G_F K_b + [D]aK_a K_c + [D]G_F K_a K_b + G_F bK_b K_c + [D]G_F abK_a K_b K_c}{K_a} \quad \text{(A71)}$$

In this equation, G_F is defined as in Equation (A63).

2.7.3 Binding of Drug to the Active Conformation of the Quaternary Complex ($DR_s{}^*GX$)

It is possible to use the microscopic constant ($abcK_c$) for the equilibrium between the ground (DR_sGX) and active ($DR_s{}^*GX$) conformations of the quaternary complex ($DRGX$) to estimate the amount of $DRGX$ complex in the active state. This calculation can be made by taking the equation for estimation of $DRGX$ [Equation (A16)] and multiplying it by the ratio $abcK_c/(abcK_c + 1)$ as shown below:

$$\frac{DR_s * GX}{R_T} = \frac{abcK_c}{abcK_c + 1} \times \frac{D}{D + K_{DRGX}} \quad \text{(A72)}$$

In this equation, K_{DRGX} is defined as described in Equation (A17).

REFERENCES

1. Jacobs, S. and Cuatrecasas, P., The mobile receptor hypothesis and "cooperativity" of hormone binding, Application to insulin, *Biochim. et. Biophys. Acta*, 433, 482–95, 1976.
2. De Lean, A., Stadel, J. M., and Lefkowitz, R. J., A ternary complex model explains the agonist-specific binding properties of the adenylate cyclase-coupled beta-adrenergic receptor, *J. Biol. Chem.*, 255, 7108–17, 1980.
3. Franklin, T. J. and Twose, P. A., Reduction in beta-adrenergic response of cultured glioma cells following depletion of intracellular GTP, *Eur. J. Biochem.*, 77, 113–7, 1977.
4. Smith, C. M., Henderson, J. F., and Baer, H. P., Effects of GTP on cyclic AMP concentrations in intact Ehrlich ascites tumor cells, *J. Cyc. Nucleo. Res.*, 3, 347–54, 1977.
5. Johnson, G. S. and Mukku, V. R., Evidence in intact cells for an involvement of GTP in the activation of adenylate cyclase, *J. Biol. Chem.*, 254, 95–100, 1979.
6. Weber, G., Energetics of ligand binding to proteins, *Adv. Prot. Chem.*, 29, 1–83, 1975.
7. Burgen, A. S., Conformational changes and drug action, *Fed. Proc.*, 40, 2723–8, 1981.
8. Ehlert, F. J., Roeske, W. R., Gee, K. W., and Yamamura, H. I., An allosteric model for benzodiazepine receptor function, *Biochem. Pharmacol.*, 32, 2375–83, 1983.
9. Ehlert, F. J., 'Inverse agonists', cooperativity and drug action at benzodiazepine receptors, *Trends Pharmacol. Sci.*, 7, 28–32, 1986.
10. Colquhoun, D., The relation between classical and cooperative models for drug action, in *Drug Receptors*, H. P. Rang, Ed., University Park, Baltimore, MD, 1973, 149–182.
11. Weiss, J. M., Morgan, P. H., Lutz, M. W., and Kenakin, T. P., The Cubic Ternary Complex Receptor-Occupancy Model, 1. Model Description, *J. Theor. Biol.*, 178, 151–167, 1996.

12. Jencks, W. P., Binding energy, specificity, and enzymic catalysis: the circe effect, *Adv. Enzymol. Mol. Biol.*, 43, 219–410, 1975.
13. Lefkowitz, R. J., Cotecchia, S., Samama, P., and Costa, T., Constitutive activity of receptors coupled to guanine nucleotide regulatory proteins, *Trends Pharmacol. Sci.*, 14, 303–307, 1993.
14. Sternweis, P. C. and Robishaw, J. D., Isolation of two proteins with high affinity for guanine nucleotides from membranes of bovine brain, *J. Biol. Chem.*, 259, 13806–13, 1984.
15. Hepler, J. R. and Gilman, A. G., G Proteins, *Trends Biochem. Sci.*, 17, 383–87, 1992.
16. Neubig, R. R., Gantzos, R. D., and Thomsen, W. J., Mechanism of agonist and antagonist binding to alpha 2 adrenergic receptors: evidence for a precoupled receptor-guanine nucleotide protein complex, *Biochemistry*, 27, 2374–84, 1988.
17. Ehlert, F. J. and Yamamura, H. I., Pharmacological analysis of muscarinic receptor-mediated inhibition of adenylate cyclase and stimulation of phosphoinositide hydrolysis, in *Molecular Mechanisms of Muscarinic Acetylcholine Receptor Function*, J. Wess, Ed., R.G. Landes Company, Austin, TX, 1995, 47–69.
18. Burgen, A. S., The effects of agonists on the components of the cardiac muscarinic receptor, *Brit. J. Pharmacol.*, 92, 327–32, 1987.
19. Wong, H. M., Sole, M. J., and Wells, J. W., Assessment of mechanistic proposals for the binding of agonists to cardiac muscarinic receptors, *Biochemistry*, 25, 6995–7008, 1986.
20. Ehlert, F. J., The relationship between muscarinic receptor occupancy and adenylate cyclase inhibition in the rabbit myocardium, *Mol. Pharmacol.*, 28, 410–21, 1985.
21. Kurose, H., Katada, T., Amano, T., and Ui, M., Specific uncoupling by islet-activating protein, pertussis toxin, of negative signal transduction via alpha-adrenergic, cholinergic, and opiate receptors in neuroblastoma × glioma hybrid cells, *J. Biol. Chem.*, 258, 4870–5, 1983.
22. Kurose, H. and Ui, M., Functional uncoupling of muscarinic receptors from adenylate cyclase in rat cardiac membranes by the active component of islet-activating protein, pertussis toxin, *J. Cycl. Nucleo. Prot. Phosphoryl. Res.*, 9, 305–318, 1983.
23. Jones, S. V., Muscarinic receptor subtypes: modulation of ion channels, *Life Sci.*, 52, 457–64, 1993.
24. Thomas, E. A. and Ehlert, F. J., Pertussis toxin blocks M2 muscarinic receptor-mediated effects on contraction and cyclic AMP in the guinea pig ileum, but not M3-mediated contractions and phosphoinositide hydrolysis, *J. Pharmacol. Exp. Ther.*, 271, 1042–50, 1994.
25. Waelbroeck, M., Gillard, M., Robberecht, P., and Christophe, J., Muscarinic receptor heterogeneity in rat central nervous system. I. Binding of four selective antagonists to three muscarinic receptor subclasses: a comparison with M2 cardiac muscarinic receptors of the C type, *Mol. Pharmacol.*, 32, 91–9, 1987.
26. Doods, H. N., Mathy, M. J., Davidesko, D., van Charldorp, K. J., de Jonge, A., and van Zwieten, P. A., Selectivity of muscarinic antagonists in radioligand and *in vivo* experiments for the putative M1, M2, and M3 receptors, *J. Pharmacol. Exp. Ther.*, 242, 257–262, 1987.
27. Hulme, E. C., Berrie, C. P., Birdsall, N. J., and Burgen, A. S., Two populations of binding sites for muscarinic antagonists in the rat heart, *Eur. J. Pharmacol.*, 73, 137–42, 1981.
28. Hilf, G. and Jakobs, K. H., Agonist-independent inhibition of G protein activation by muscarinic acetylcholine receptor antagonists in cardiac membranes, *Eur. J. Pharmacol.*, 225, 245–52, 1992.

29. Ehlert, F. J., Allosterism and drug action, *Proc. West. Pharmacol. Soc.*, 35, 215–9, 1992.
30. Onaran, H. O. and Costa, T., Agonist efficacy and allosteric models of receptor action, *Ann. New York Acad. Sci.*, 812, 98–115, 1997.
31. Higashijima, T., Ferguson, K. M., Sternweis, P. C., Ross, E. M., Smigel, M. D., and Gilman, A. G., The effect of activating ligands on the intrinsic fluorescence of guanine nucleotide-binding regulatory proteins, *J. Biol. Chem.*, 262, 757–61, 1987.
32. Higashijima, T., Ferguson, K. M., Sternweis, P. C., Ross, E. M., Smigel, M. D., and Gilman, A. G., The effect of GTP and Mg2+ on the GTPase activity and the fluorescent properties of Go, *J. Biol. Chem.*, 262, 757–61, 1987.
33. Waelbroeck, M., Boufrahi, L., and Swillens, S., Seven helix receptors are enzymes catalysing G protein activation. What is the agonist Kact?, *J. Theor. Biol.*, 187, 15–37, 1997.
34. Berrie, C. P., Birdsall, N. J., Burgen, A. S., and Hulme, E. C., Guanine nucleotides modulate muscarinic receptor binding in the heart, *Biochem. Biophys. Res. Comm.*, 87, 1000–5, 1979.
35. Haga, K., Haga, T., and Ichiyama, A., Reconstitution of the muscarinic acetylcholine receptor, Guanine nucleotide-sensitive high affinity binding of agonists to purified muscarinic receptors reconstituted with GTP-binding proteins (Gi and Go), *J. Biol. Chem.*, 261, 10133–40, 1986.
36. Ehlert, F. J. and Rathbun, B. E., Signaling through the muscarinic receptor-adenylate cyclase system of the heart is buffered against GTP over a range of concentrations, *Mol. Pharmacol.*, 38, 148–58, 1990.
37. Lazareno, S., Measurement of agonist-stimulated [^{35}S]GTP gamma S binding to cell membranes, *Meth. Mol. Biol.*, 106, 231–45, 1999.
38. Ehlert, F. J., Correlation between the binding parameters of muscarinic agonists and their inhibition of adenylate cyclase activity, *Adv. Exp. Med. Biol.*, 236, 265–76, 1988.
39. Watson, M., Yamamura, H. I., and Roeske, W. R., [^{3}H]pirenzepine and (–)-[^{3}H]quinuclidinyl benzilate binding to rat cerebral cortical and cardiac muscarinic cholinergic sites. I. Characterization and regulation of agonist binding to putative muscarinic subtypes, *J. Pharmacol. Exp. Ther.*, 237, 411–18, 1986.
40. Ek, B. and Nahorski, S., Muscarinic receptor coupling to inositol phospholipid metabolism in guinea-pig cerebral cortex, parotid gland and ileal smooth muscle, *Biochem. Pharmacol.*, 37, 4461–7, 1988.
41. Wess, J., Bonner, T. I., Dorje, F., and Brann, M. R., Delineation of muscarinic receptor domains conferring selectivity of coupling to guanine nucleotide-binding proteins and second messengers, *Mol. Pharmacol.*, 38, 517–23, 1990.
42. Savarese, T. M., Wang, C. D., and Fraser, C. M., Site-directed mutagenesis of the rat m1 muscarinic acetylcholine receptor. Role of conserved cysteines in receptor function, *J. Biol. Chem.*, 267, 11439–48, 1992.
43. Caulfield, M. P., Muscarinic receptor classification, in *Muscarinic Receptor Subtypes in Smooth Muscle*, R. M. Eglen, Ed., CRC Press, Boca Raton, FL, 1997, 1–37.
44. Colquhoun, D., Affinity, efficacy and receptor classification, is the classical theory still useful?, in *Perspectives on Receptor Classification*, J. Black, D. Jenkinson, and V. Gerskowitch, Eds., Alan R. Liss, New York, 1987, 103.
45. Stephenson, R. P., A modification of receptor theory, *Brit. J. Pharmacol.*, 11, 379–93, 1956.
46. Furchgott, R. F., The use of β-haloalkylamines in the differentiation of receptors and in the determination of dissociation constants of receptor-agonist complexes, *Adv. Drug Res.*, 3, 21–55, 1966.

47. Furchgott, R. F. and Bursztyn, P., Comparison of dissociation constants and of relative efficacies of selected agonists acting on parasympathetic receptors, *Ann. New York Acad. Sci.*, 144, 882–99, 1967.
48. Waelbroeck, M., Kinetics versus equilibrium: the importance of GTP in GPCR activation, *Trends Pharmacol Sci.*, 20, 477–81, 1999.
49. Mackay, D., An analysis of functional antagonism and synergism, *Brit. J. Pharmacol.*, 73, 127–34, 1981.
50. Kenakin, T. P. and Beek, D., *In vitro* studies on the cardiac activity of prenalterol with reference to use in congestive heart failure, *J. Pharmacol. Exp. Ther.*, 220, 77–85, 1982.
51. Black, J. W. and Leff, P., Operational models of pharmacological agonism, *Proc. Roy. Soc. Lond. B*, 220, 141–62, 1983.
52. Ehlert, F. J., Griffin, M. T., Sawyer, G. W., and Bailon, R., A simple method for estimation of agonist activity at receptor subtypes: comparison of, *J. Pharmacol. Exp. Ther.*, 289, 981–92, 1999.
53. Black, J. W., Leff, P., Shankley, N. P., and Wood, J., An operational model of pharmacological agonism: the effect of E/[A] curve shape on agonist dissociation constant estimation, *Brit. J. Pharmacol.*, 84, 561–71, 1985.
54. Ehlert, F. J., Coupling of muscarinic receptors to adenylate cyclase in the rabbit myocardium: effects of receptor inactivation, *J. Pharmacol. Exp. Ther.*, 240, 23–30, 1987.
55. Kellett, E., Carr, I. C., and Milligan, G., Regulation of G protein activation and effector modulation by fusion proteins between the human 5-hydroxytryptamine(1A) receptor and the alpha subunit of G(i1): differences in receptor-constitutive activity imparted by single amino acid substitutions in G(i1)alpha, *Mol. Pharmacol.*, 56, 684–92, 1999.
56. Trzeciakowski, J. P., Stimulus amplification, efficacy, and the operational model. Part I — Binary complex occupancy mechanisms, *J. Theor. Biol.*, 198, 329–46, 1999.
57. Colquhoun, D., Binding, gating, affinity and efficacy: the interpretation of structure-activity relationships for agonists and of the effects of mutating receptors, *Brit. J. Pharmacol.*, 125, 924–947, 1998.
58. Mei, L., Lai, J., Yamamura, H. I., and Roeske, W. R., The relationship between agonist states of the M1 muscarinic receptor and the hydrolysis of inositol lipids in transfected murine fibroblast cells (B82) expressing different receptor densities, *J. Pharmacol. Exp. Ther.*, 251, 90–7, 1989.
59. Lu, Z. L. and Hulme, E. C., The functional topography of transmembrane domain 3 of the M-1 muscarinic acetylcholine receptor, revealed by scanning mutagenesis, *J. Biol. Chem.*, 274, 7309–15, 1999.
60. Kenakin, T., Agonist-receptor efficacy. I: Mechanisms of efficacy and receptor promiscuity [see comments], *Trends Pharmacol Sci.*, 16, 188–92, 1995.
61. Leff, P., Scaramellini, C., Law, C. and McKechnie, K., A three-state receptor model of agonist action, *Trends Pharmacol. Sci.*, 18, 355–62, 1997.
62. Jordan, B. A. and Devi, L. A., G protein-coupled receptor heterodimerization modulates receptor function, *Nature*, 399, 697–700, 1999.
63. Furchgott, R. F., Pharmacological characterization of receptors: its relation to radioligand-binding studies, *Fed. Proc.*, 37, 115–20, 1978.
64. del Castillo, J. and Katz, B., Interaction at end-plate receptors between different choline derivatives, *Proc. Roy. Soc. Lond. Series B*, 146, 369, 1957.
65. Dawkins, R., *The Extended Phenotype: The Long Reach of the Gene*, Oxford University Press, Oxford, New York, 1999.

3 From the Static to the Dynamic: Three Models of Signal Transduction in G Protein-Coupled Receptors

Peter J. Woolf and Jennifer J. Linderman

CONTENTS

> Within the next half century, with all genes identified and all possible cellular interactions and reactions charted, pharmacologists developing a drug or toxicologists trying to predict whether a substance is poisonous may well turn to computer models of cells to answer their questions.
>
> — Francis Collins, 1999[1]

3.1 INTRODUCTION

Ligand binding to cell surface receptors initiates a series of events known as signal transduction. In this process, receptors alter the states of other proteins, which in turn leads to an avalanche of changes inside the cell. The cell uses this avalanche of information to describe what is occurring in the cellular environment and then alters its behavior accordingly. Without signal transduction, the cell would be blind to many environmental changes, making the cellular cooperation we see in most organisms

0-8493-0100-9/01/$0.00+$.50

impossible. Thus understanding signal transduction is key to understanding how cells communicate with each other and how we can communicate with cells as well.

Much of pharmacology research is focused on developing drugs to manipulate the outputs of signal transduction pathways. For example, a drug acting on a receptor could cause a cell to grow, differentiate, migrate, contract, or secrete some other compound. Unfortunately, signal transduction is not a simple process; signaling involves a complex network of intracellular and cell surface reactions. Modern molecular biologists have done an excellent job of identifying the participants of this complex network and even many of the connections between participants. However, before we can accurately predict how a drug will act on a cell, we need to determine how all of these interconnected pieces behave when the system is reassembled.

We believe that our understanding of signal transduction will rest on accurately evaluating the influence of *quantitative* factors such as concentrations, reaction rate constants, and multiple reaction pathways on the signaling behavior. By gathering quantitative data on signal transduction, we can compare experimental results to theoretical predictions made by mathematical models such as those presented in this chapter. By developing accurate mathematical models, we can gain insight into signaling events that may not be experimentally accessible, either due to limited resolution of an assay or unavailability of the technology required to observe some events. These models may also suggest new hypotheses to be tested.

Most modeling in pharmacology focuses on predicting drug action as it relates to a particular response. Many early and still useful models consist of essentially phenomenological descriptions of drug action, such as efficacy and potency, that can quantify drug action but are difficult to relate to the physical events that cause the drug response. Here the process of signal transduction is essentially a black box. Some newer models, which will be described in this chapter, are mechanistic, meaning that these models explicitly take into account the cell biology of signal transduction. These mechanistic models have an advantage in that they can be used to predict how a signal transduction pathway will respond to a drug with a particular set of properties. Although the constants needed to describe a mechanistic model are more difficult to obtain than those for the phenomenological models, much of the required information may one day be available using techniques from modern molecular biology.

3.1.1 G Protein-Coupled Receptors

In this chapter, we focus on an important class of receptors that transmit their signal by coupling with guanine nucleotide binding proteins (G proteins). Approximately 80% of all known hormones, neurotransmitters, and neuromodulators are believed to transmit their signal through these G protein-coupled receptors (GPCRs),[2] and pharmacologists estimate that 60% of all medicines used today act through G protein signaling pathways.[3] Examples of such GPCR systems include the β-adrenergic receptor (involved in regulating heart contractility), the opioid and dopamine receptors (involved in brain function), and the N-formyl peptide receptor (involved in the immune response). Abnormalities in GPCR signaling are involved in numerous diseases and disorders and are therefore a major target for therapeutic intervention. By working to understand GPCR signaling, we hope to foster the development of better, more effective drugs to improve human health.

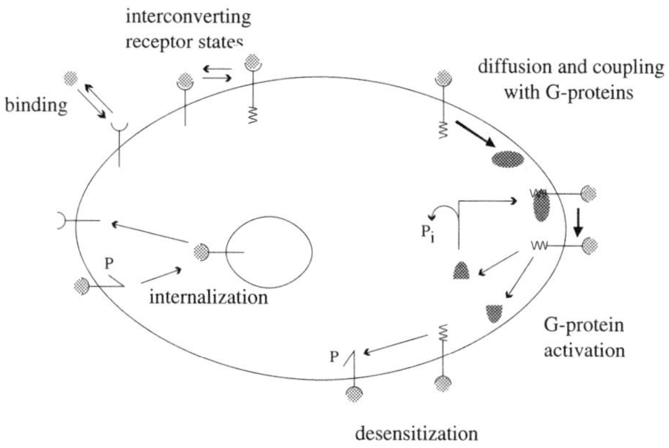

FIGURE 3.1 Events involved in G protein-coupled receptor signaling. Ligands not only influence how the receptor transmits its signal to the G protein, but could also affect the receptor's rate of desensitization and internalization, along with a number of other less well understood effects.[9,24]

The G protein activation cycle is summarized in many reviews[4,5] and will only be briefly described here. The process begins when active receptors bind inactive G proteins, the heterotrimer αβγ-GDP. In the presence of GTP, the G protein then exchanges GTP for GDP on its α subunit, causing the dissociation of the G protein into two subunits: α–GTP and βγ. Both α–GTP and βγ are believed to propagate the signal inside the cell, thereby initiating the changes that eventually lead to a response within seconds to minutes of receptor activation. The α subunit possesses intrinsic GTPase activity that deactivates the α subunit, allowing the α–GDP and βγ subunits to reform the inactive G protein. Thus G protein activation serves as a timed molecular switch. After sustained receptor activation, the receptor itself is also regulated in a process termed desensitization. Desensitization may play a key role in processes such as drug addiction, drug tolerance, and cellular development.[6,7]

Signaling in GPCRs is controlled by many interrelated processes that govern not only what signals are detected, but also how long those signals persist. Some of these processes are diagrammed in Figure 3.1. Receptors (1) bind and dissociate extracellular ligand, (2) interconvert between inactive and active conformations, terms that refer to their ability to activate G proteins, (3) diffuse and couple with G proteins, (4) transmit signals via G protein activation, (5) desensitize at the cell surface, and (6) internalize and may recycle. The length of time a G protein remains active depends on its GTPase activity, which may be influenced by another class of proteins termed RGS ("regulators of G protein signaling") proteins.[5,8] This list is not meant to be all inclusive, and these receptors (also known as seven transmembrane receptors) are now also believed to be involved in signal transduction events distinct from G protein activation.[9]

A unique property of GPCRs is that different drugs can induce a wide variety of responses ranging from strong positive responses (elicited by positive agonists) to strong negative responses (elicited by inverse agonists) as shown in Figure 3.2.

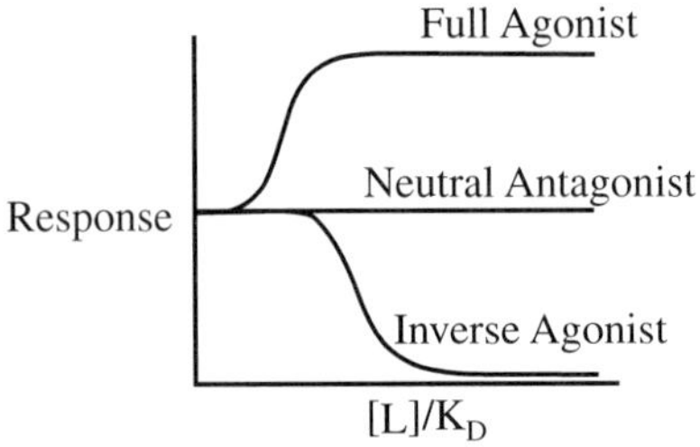

FIGURE 3.2 Three classes of ligands for GPCRs.

Positive agonists are thought to induce or select the active (R*) state of the receptor, which rapidly catalyzes the GDP/GTP exchange on the G protein. In contrast, inverse agonists induce or select the inactive receptor state (R), driving the signal below the basal (constitutive) level. In this way, the activity of a drug acting on a GPCR depends on the ability of that drug to both bind to the receptor and to change the receptor conformation. Other physical factors — for example, the concentrations of various types of molecules involved in the pathways or the diffusivity of molecules in the membrane — also play roles in determining the response levels and possibly direction. Models of GPCR action can help us to both discover and describe the impact of such factors.

3.1.2 Types of Models

In this chapter, we discuss three mechanistic models of GPCR activation and illustrate how each model can yield useful insights. A majority of mechanistic models can be categorized as equilibrium, steady state, or unsteady state models. Neutral antagonists do not influence the receptor conformation, but block the effects of other ligands. Equilibrium models deal with systems that do not change with time and require no energy to sustain. Steady state systems also do not change with time, but require some sort of energy input to maintain. Unsteady state models describe systems that do change with time.

To illustrate the difference between these various types of models, consider the physical example of a pendulum shown in Figure 3.3. The equilibrium solution of a pendulum places the pendulum hanging straight down. This is the state that the pendulum is "trying" to get to, but keeps swinging past. If the pendulum were instead driven in one direction by a motor or a constant breeze, then we would use a steady state model, which would place the pendulum at an angle. This new state is similar to the equilibrium state except that it includes the effects of some external energy input (drag from the wind or electricity for the motor). Finally, to look at the time-varying swinging of the pendulum we can use an unsteady state model. The unsteady state model will show us how high the pendulum swings, how fast the pendulum moves, and how long it will take the pendulum to stop moving. Thus our choice of which model to use depends on the type of behavior we want to describe.

Equilibrium models are the most commonly used type of model in pharmacology. One of the main reasons for this model's popularity is that this class of model tends

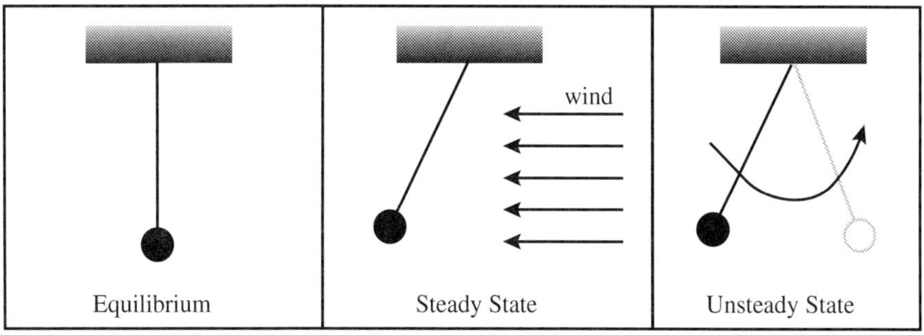

FIGURE 3.3 A pendulum analogy to describe the three types of models discussed in this chapter. At equilibrium, the pendulum hangs straight down. When the pendulum is driven by an energy source (i.e., wind), the pendulum hangs at an angle in a steady state. The unsteady state model describes how the pendulum moves with time.

to have parameters that are easily measured and the mathematics are generally simple. By definition, these models assume that the system is at thermodynamic equilibrium — a state that requires no energy to maintain and does not change with time.

In biology, the equilibrium state corresponds to death; however, equilibrium models can still be useful for modeling certain pharmacological systems. For example, some experimental systems such as membrane preparations can be used to obtain equilibrium data. Although these *in vitro* measurements may not correspond exactly to the *in vivo* situation, in many cases the agreement is sufficiently good to warrant using this simpler system. In other cases, equilibrium models can be used to describe an "apparent" equilibrium between species in the cell, even though the system might in truth be an energy-requiring steady state system. For example, consider the "equilibrium" number of receptors on the cell surface. In truth, receptors are constantly being shuttled to and from the cell surface. However, at any given time there appears to be a fixed number of receptors on the surface. In this way, equilibrium models are often used as a first approximation to fit the mechanics of the system.

Steady state models describe a system that is not in equilibrium, but the apparent concentrations of each species do not change. For example, consider a metabolic cycle like the Kreb's cycle, which converts acetyl CoA to energy and CO_2. The concentrations of intermediates in the cycle do not change with time; however, there is a constant flux of acetyl CoA into the system. To maintain such a state of stable disequilibrium, the system requires a constant input of energy in some form, in this case the bond energy of acetyl CoA.

Steady state models are generally more realistic models of biological systems than equilibrium models. Nearly every biological system requires energy to maintain its state, and as such, nearly every biological reaction is in a constant state of flux. For models of drug action, ongoing and energy-requiring processes such as G protein activation and receptor phosphorylation can have profound effects on the behavior of a drug and are important to include.

Unsteady state models provide an even more general description of a system because they include the effects of time. Steady state and equilibrium models assume that the system has had an infinite amount of time to stabilize, and as such remove all dependence on time and initial conditions. In contrast, unsteady state models describe how the system evolves with time and can be used to describe systems that have no steady state, such as periodic or chaotic systems.

Unfortunately, unsteady state models are also some of the most complicated to solve. Often, these models are formulated as coupled sets of differential equations and in many cases there is no analytical solution. Even a numerical solution may be difficult to obtain. In spite of these problems, unsteady state models are probably the best choice if sufficient data are available to motivate their development.

One disadvantage to using the more detailed models, e.g., unsteady state models, is the need for more parameters to describe the system. For example, for the equilibrium solution to the pendulum problem we do not need any information because the pendulum is always down no matter how much it weighs or how long it is. In contrast, for the unsteady state solution, we must know all about the pendulum plus where the pendulum was started and how fast it was moving when it started. In complicated biological systems such as a signal transduction pathway or metabolic cycle, a large number of parameters are also needed to describe the unsteady state behavior. However, by using many parameters to describe a complex biological system we are not "fitting an elephant." This is because the parameters in the unsteady state model describe specific physical phenomena related to the mechanics of the process, not empirical correction factors used to simply *fit* the data. The parameters used in describing these biological systems are (at least in principle) measurable physical quantities just like the mass and length used to describe the pendulum. Fortunately, in many cases the exact parameter values are not needed because we are only interested in the qualitative behavior of the system. For example, we may be interested in determining whether a ligand is a positive or inverse agonist, and whether it is possible for a ligand to be both, but under different circumstances. Thus, even crude mathematical models can be helpful in predicting the behavior of the response, even though they may not give the exact magnitude or duration of the response.

3.1.3 Species Conservation

Along with the type of model we choose, we must also determine which simplifying assumptions are reasonable to make. One major simplifying assumption that is often made in pharmacological models is that the concentration of some species is constant. For example, during an equilibrium binding experiment, the free ligand concentration may not change significantly. In this case, we can approximate the system by modeling the free ligand concentration as a constant, thereby ignoring any dynamics that might come about due to ligand depletion. The concentration of a species can be approximated as a constant if (1) the species is in great stoichiometric excess or (2) the system has access to a near infinite reservoir of the species at a fixed concentration.

When these conditions are not met, however, depletion effects must be included through the use of species conservation equations. For example, receptors in an equilibrium binding experiment are known to saturate (depletion of free receptors), implying that we need to include a receptor conservation equation in our model. This conservation equation can be stated as:

$$R_{bound} + R_{free} = R_{total} \tag{1}$$

This species conservation equation enforces the fact that the model has access to only R_{total} receptors, making the maximum number of receptors that can bind ligand equal to the total number of receptors.

Why isn't species conservation always included? We know that no species *really* comes from an infinite source, so we are always safe in including species conservation equations even if they have little effect on the predictions of the model. However, including species conservation also makes the model equations more difficult to solve, as the model often changes from linear to nonlinear. Linear models are generally easy to solve both analytically and numerically, and the results they generate are easy to interpret. In a linear model, doubling the value of an input will always double the output. In contrast, nonlinear models often have no analytical solution and can be more difficult to solve numerically. Interpretation of nonlinear models can also be more challenging because, for example, doubling the value of an input may or may not double the output. Thus many researchers choose not to include species conservation even if it is known to play an important role, so as to simplify their models enough to tease out some, but perhaps not all, relevant behaviors.

We believe that unsteady state models that include species conservation will ultimately provide the best insight into GPCR signal transduction. Signal transduction via GPCRs is an inherently unsteady state process because it is designed to detect *changes* in the environment — time varying information. Thus GPCR pharmacology is essentially the study of how to manipulate the information given to cells to change their states, making unsteady state models an invaluable tool in this field. For an unsteady state model to function properly we must also allow the concentration of each species (e.g., free receptors, inactive G proteins) to vary, and if need be deplete, and thus species conservation must be included. However, this is not to say that equilibrium models or models without species conservation are useless. On the contrary, equilibrium models describe where the system "wants" to go given enough time, and as such can be used as a landmark when analyzing the more complicated unsteady state systems. Models without species conservation describe a system under a limited set of conditions, but in a much simpler way. Therefore, it is instructive to begin by looking at simple equilibrium models and then see how these models can help us understand more complicated unsteady state models. Through the use of simpler models to understand more complex and realistic models, we can begin to explain and predict the complex events of GPCR signaling, giving us insight into the mechanisms of drug action on this receptor.

3.2 EQUILIBRIUM MODELS

In this model section and in the two that follow, we begin first with a simple didactic model before moving on to the case of GPCRs. A simple example of an equilibrium

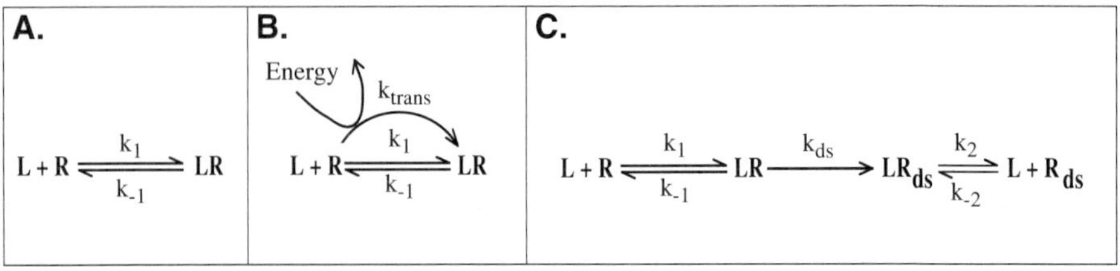

FIGURE 3.4 Three representative models of receptor-ligand binding: (a) equilibrium; (b) steady state; and (c) unsteady state. These are simple models used to illustrate the principles behind each technique.

process is receptor-ligand binding (Figure 3.4a). In this model, ligands bind to receptors at a rate proportional to k_1 (M^{-1} sec^{-1}) and dissociate with rate constant k_{-1} (sec^{-1}). At equilibrium, the forward reaction is exactly balanced by the reverse reaction, resulting in the equilibrium relationship:

$$K_{eq} = \frac{k_{-1}}{k_1} = \frac{[R][L]}{[RL]} \tag{2}$$

Here [R] represents the concentration of free receptors on the cell surface, [RL] is the number of ligand-bound receptors, [L] is the free ligand concentration, and K_{eq} is the equilibrium dissociation constant. In the equilibrium state there is no net binding or dissociation of the ligand from the receptor, so the system is at thermodynamic equilibrium — the state that receptors and ligands "want" to go to if left to themselves. To include receptor conservation, the term [R] can be replaced with R_{tot} – [RL] to give the familiar expression for receptor-ligand binding:

$$K_{eq} = \frac{k_{-1}}{k_1} = \frac{(R_{tot} - [RL])[L]}{[RL]} \tag{3}$$

By knowing the free ligand concentration (generally assumed equal to the ligand concentration added, a reasonable assumption if the number of ligand molecules is significantly greater than the number of receptor molecules[10]), the number of ligand-bound receptors, and total number of receptors, we can calculate the equilibrium constant between the bound and free receptors.

Although signal transduction via GPCRs begins with receptor-ligand binding, binding alone is insufficient to describe the ligand's effects on the cell. For example, receptor-ligand binding alone cannot predict inverse agonism because this would require the receptors to be bound with a *negative* number of ligands. Clearly other signaling phenomena must be included. In the case of GPCRs the natural next step is to include more of the signal transduction pathway, such as the G proteins that receive the signal inside the cell.

What happens to the predicted cellular response when G proteins are added? G protein reactions, such as receptor-G protein binding, allow us to simulate the first

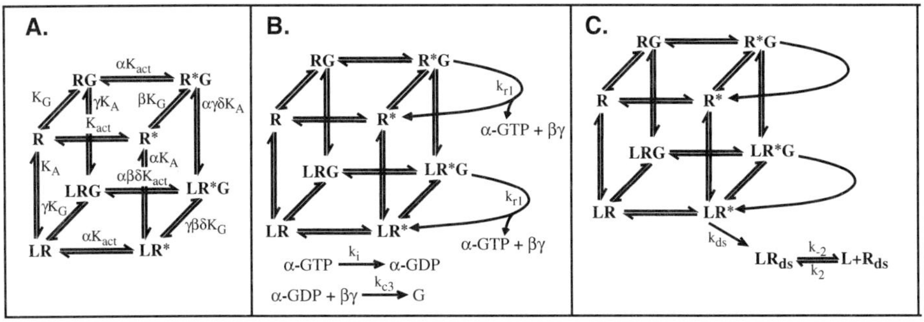

FIGURE 3.5 Three models of signal transduction via GPCRs. (a) The Cubic Ternary Complex (CTC) model[15] that describes the equilibrium between receptor, G protein, and ligand. (b) A steady state variant of the CTC model that includes an explicit G protein activation step.[23,24] (c) An unsteady state version of the CTC model that includes a desensitization step for ligand-bound active receptors.

signaling events inside the cell, thereby generating a more accurate picture of the signal transduction events caused by a particular ligand. Recent evidence suggests that GPCRs exist in at least two signaling-capable states on the cell surface, an "active" or R* conformation and an "inactive" or R conformation.[11,12] The R* state is able to interact with G proteins to produce G protein activation; the R state is not. These active and inactive states presumably represent sets of conformations of the receptor,[13] and the receptors are likely to move between these states fairly rapidly.

GPCRs are typically modeled using what are termed ternary (for ligand/receptor/G protein) complex models, or TCMs (reviewed in Reference 14). Recent models that include the R* and R states are the extended ternary complex model[11] and the more thermodynamically complete cubic ternary complex (CTC) model.[15,16] The CTC model is shown in Figure 3.5a. These models (and other related TCMs) are equilibrium models and have found some use in describing unique drug properties.

In the CTC model shown in Figure 3.5a, ligands influence the distribution of R* and R states (and thus potentially efficacy) via the *conformational selectivity parameter,* α. At high α values ($\alpha > 1$) the ligand induces or selects the active receptor state (positive agonism), whereas when α is low ($\alpha < 1$) the ligand induces or selects the inactive receptor state (inverse agonism). When α is exactly equal to 1, the ligand has no effect on the receptor conformation and does nothing but block the effects of other drugs (neutral antagonism). In the CTC model, α is just one of many ligand-specific parameters that determines how the cell will respond to a drug.[15]

One unexpected ligand property that the CTC model predicts is the ability of a ligand to change from a positive agonist to a negative agonist (or vice versa) — something that might be termed *agonist inversion.* In agonist inversion, the efficacy of the ligand can be modulated simply by changing a parameter in the model, such as the conformational selectivity α (a ligand-specific parameter) or the G protein concentration (reflecting a state inside the cell). This phenomenon of agonist inversion is similar to "protean agonism" introduced by Kenakin[17,18]; however, protean agonism only focuses on changing K_{act}, while agonist inversion describes a more general phenomenon that could be caused by changes in other parameters.

Using the CTC model, we recently found a simple expression that can be used to predict whether positive or inverse agonism is expected for a particular set of parameters. To do this, we calculated the ratio between the signal produced by receptors saturated with ligand to receptors without bound ligand to be:

$$\alpha\delta\gamma + GK_{G^{\gamma}}(\alpha\delta - 1) + K_{act}\alpha(\delta\gamma - 1)\begin{cases} > 1 \text{ positive agonism} \\ = 1 \text{ neutral agonism} \\ < 1 \text{ inverse agonism} \end{cases} \quad (4)$$

where α, δ, γ, K_G, K_{act} are equilibrium parameters from Figure 3.5a and G is the number of free G proteins (in the GDP-bound form, i.e., "activate-able" G proteins). If this ratio is equal to one, then the signal is the same with or without ligand; thus the ligand behaves as a neutral antagonist. If the ratio is greater than one, the ligand acts as a positive agonist. Similarly, when the ratio is less than one, the ligand acts as an inverse agonist.

Equation (4) was found with the assumption of excess ligand (often reasonable) and a constant concentration of free G protein (G). The latter is probably not a physically realistic assumption. In the real system, active receptors transform free G proteins into a form that can no longer bind to receptors. Receptor activation should significantly reduce the free G protein concentration available to the receptor. Unfortunately, the process of G protein activation cannot be described by an equilibrium model, and as such is not described in Equation (4). We have found an expression analogous to Equation (4) when G protein conservation is included in the model; however, this expression is significantly more cumbersome and does not add any qualitatively new information. Thus, ignoring species conservation (here, G protein conservation) is useful in this situation for it allows us to present a small, analytical expression that captures the qualitative behavior of the system.

Notice that Equation (4) can have both positive and negative terms. By varying the values of the parameters, one can obtain values that range from well below 1 to well above 1 — thereby spanning the range from inverse agonist to positive agonist. For example, by varying K_{act}, we obtain the protean agonism behavior described in the literature.[17,18] Interestingly, varying the G protein number can also cause agonist inversion, implying a novel mechanism for the cell to regulate the information it receives. Cells are thought to regulate the number of G proteins they express [2]. For example, exposing human neutrophils to N-formyl peptide ligands has been shown to upregulate the number of G proteins,[19,20] although the magnitude and timing of this increase have not been carefully studied.

According to Equation (4), this change in G protein number could have a profound effect on how the cell responds to a ligand. For example, imagine that a cell is stimulated by a positive agonist. By definition, a positive agonist means that the ligand-stimulated signal is greater than the unstimulated signal and Equation (4) has a value greater than 1. However, if the cell reduces the number of G proteins able to interact with the receptor (by degradation or sequestration for example), then the value of the second term in Equation (4) would drop. Thus the cell could selectively tune out that particular ligand, changing it from a strong positive agonist

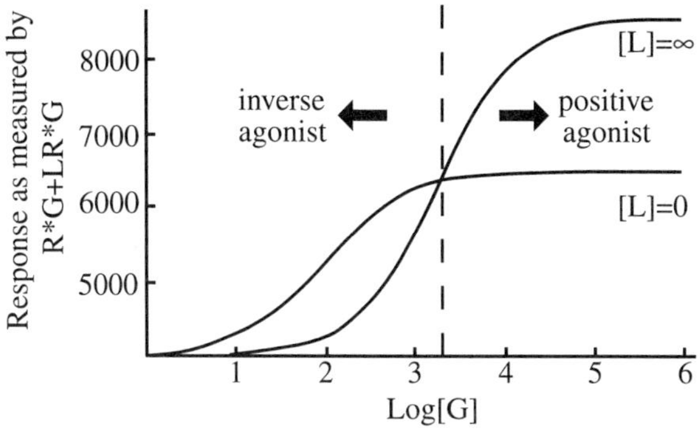

FIGURE 3.6 An example of agonist inversion. The two curves describe the effects of G protein concentration on the cellular response when ligand concentration is zero (no ligand present) and infinite (receptors are saturated with ligand). At low G protein concentration, the constitutive signaling ([L] = 0) is greater than the signal when ligand is present at saturating concentrations, and thus the ligand behaves as an inverse agonist. In contrast, at high G protein concentrations, the receptors saturated with ligand have a greater signal than at the constitutive level, thus the ligand behaves as a positive agonist.

to a neutral antagonist or inverse agonist, effectively ignoring that ligand. Note that while the cell may tune out one ligand, it may still be able to detect other ligands that have different parameters (e.g., α).

Agonist inversion caused by changes in G protein number is demonstrated in the simulation results shown in Figure 3.6. At a low free G protein concentration, the signal without ligand is greater than the signal with ligand, and the ligand behaves as an inverse agonist. However, once the free G protein concentration reaches a critical value, the ligand begins to behave as a positive agonist. Therefore, under conditions for agonist inversion, a positive agonist does not have to always be a positive agonist, but instead could vary from cell to cell depending on the state of the signal transduction pathway. Thus we must consider both the drug's impact on the cell as well as the cell's impact on the drug when describing drug efficacy.

Qualitatively, agonist inversion can be thought of as a transition between two qualitatively different states. Take as an analogy the two possible states of a pendulum: swinging back and forth or spinning around in circles, like the propeller of an airplane. Depending on how the pendulum was pushed, how heavy the weight, and how long the arm, the pendulum will either swing back and forth or will spin. Similarly, depending on the equilibrium constants of the CTC model, most drugs will either be a positive agonist or negative agonist. By changing the values of the equilibrium constants, we can then get a qualitatively different behavior, just as in the case of a pendulum we can change a spinning pendulum into a swinging pendulum just by increasing the weight. However, a living cell strongly differs from a pendulum in that the cell can dynamically change its parameters in response to its environment.

It is not known whether agonist inversion occurs in nature. Chidiac et al.[21] found that dichloroisoproterenol acting on β_2-adrenergic receptors could show either positive or inverse agonist behavior depending upon the cell preparation used in the experiment. One possible explanation is that the G protein number in each cell preparation was different, causing an agonist inversion in some cells as described by Equation (4). Perhaps the reason that there is relatively little experimental support for agonist inversion is because we just have not done the right experiment, or don't have the sensitivity to detect the small changes involved. Similarly, although agonist inversion is predicted by an equilibrium version of the CTC model, maybe other nonequilibrium effects such as receptor desensitization or G protein activation also affect inversion. It is worthwhile to note that agonist inversion would be very difficult to understand without a theoretical understanding of this signal transduction pathway, while at the same time the theoretical understanding is built on experimental evidence.

3.3 STEADY STATE MODELS

To illustrate how a steady state model could evolve before examining such a model for GPCRs, consider the receptor-ligand binding as described in Figure 3.4b. In this scenario, imagine that the ligand is actively transported toward the receptor, thereby introducing an energy-requiring step that would disrupt the equilibrium. In our model, this active transport could be included by adding an additional binding step governed by the rate constant k_{trans} (M^{-1} sec^{-1}). The overall rates of formation of R and LR can be written as:

$$\begin{aligned} r_R &= k_{-1}[LR] - k_1[L][R] - k_{trans}[L][R] \\ r_{LR} &= k_1[L][R] + k_{trans}[L][R] - k_{-1}[LR] \end{aligned} \tag{5}$$

By setting the rates of formation equal to zero, we can obtain an expression for the steady state values of [R], [L], and [LR]:

$$K^*_{eq} = \frac{k_{-1}}{k_1 + k_{trans}} = \frac{[R][L]}{[RL]} \tag{6}$$

Note the similarity to the equilibrium model solution [Equation (2)]. Here the apparent equilibrium constant, K^*_{eq}, can be related to the true equilibrium constant by adding k_{trans} to the denominator. Another way to interpret this steady state result is that the form of the steady state solution is identical to the equilibrium solution except that the value of the equilibrium constant is different. This relationship between equilibrium and steady state models is true as long as all steps are reversible. When irreversible steps are included, the structure of the solutions may also change. The implication of this finding is that experimental data taken from true equilibrium systems (such as membrane extracts) cannot easily be applied to living steady state systems. In the example given here, the equilibrium dissociation constant K_{eq} could be much greater than the equilibrium dissociation constant K^*_{eq} in the steady state system, causing a significant error in the interpretation of *in vivo* data based on data from membrane systems. Adams et al.[22] describe a similar example in which the amount of receptor/G protein precoupling (association prior to ligand binding) measured in membrane assays may overestimate that present *in vivo*.

A steady state analysis can be done on more complicated systems such as GPCR activation of G proteins. As discussed earlier, G proteins cycle from an active GTP-bound state to an inactive GDP-bound state. In modeling terms, this cycling from an active to an inactive state indicates that this process cannot be modeled as an equilibrium process because it requires energy to maintain. By adding an explicit G protein activation step to the CTC model, we create a steady state model that follows the signal transduction pathway farther into the cell. This model is shown in Figure 3.5b.[23,24]

This new model of GPCR activation has many of the same properties of the equilibrium CTC model. However, as described above, the values of many of the parameters can be expected to be different. For example, consider the reversible reaction between R* and R*G shown in Figure 3.7. The analysis of this step is identical to that done earlier for receptor-ligand binding shown in Figure 3.4b, resulting in the following apparent equilibrium constant:

$$K^*_G = \frac{k_2}{k_{-2} + k_{r1}} = \frac{[R^*G]}{[R^*G][G]} \tag{7}$$

Thus if k_{r1} is set to zero, then we recover the original equilibrium model.

Note that in this steady state model, we must explicitly keep track of the number of inactive and active G proteins by using a species conservation relationship for the model to be realistic. Before receptor activation, very few of the G proteins should be in the active state (in a system with low constitutive activity), and therefore a majority of the G proteins will be in the free, inactive state (αβγ-GDP, denoted as simply G in the model schematic and equations). However, upon agonist binding, the receptor should activate many of the accessible G proteins in order to propagate the signal inside the cell. Therefore, the free G protein concentration [G] must change as a key step in the dynamics of signal transduction.

A striking example of the difference between the equilibrium and steady state models is shown in Figure 3.8 for one set of reasonable parameter values. Here the equilibrium model predicts the ligand will behave as a positive agonist, while the steady state model predicts inverse agonism. Why are the predictions so different? There are two contributing factors. First, as noted before, many steady state models

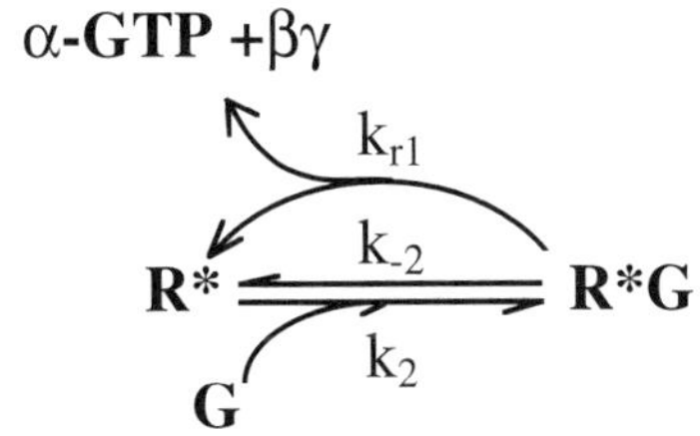

FIGURE 3.7 A steady state G protein activation reaction. This reaction is just one step in the steady state CTC model of Figure 3.5b.

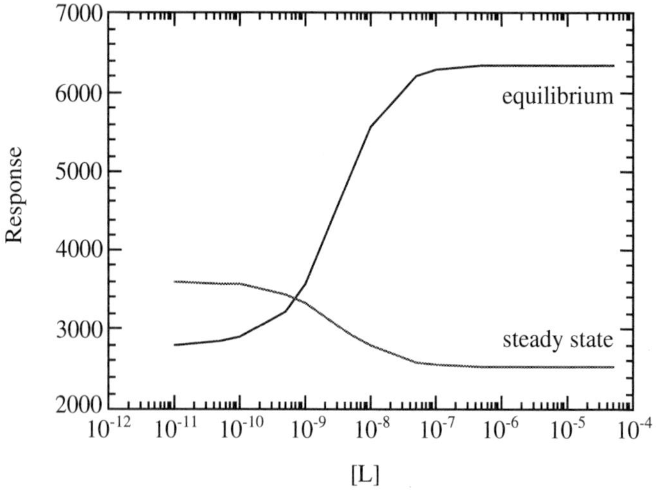

FIGURE 3.8 Responses predicted with the equilibrium and steady state version of the Cubic Ternary Complex model. For a particular set of realistic parameter values, the equilibrium model predicts that the drug will behave as a positive agonist, while the steady state model predicts that the drug will be an inverse agonist.

are similar to equilibrium models but with different parameter values. In fact, any steady state model should give the equilibrium solution when no energy is put into the system (no GTP present in the case of GPCR activation). In other words, steady state models are more general forms of equilibrium models. In our earlier analysis we showed that the equilibrium solutions of the CTC model could show either positive agonist or inverse agonist behavior, depending on the values of the parameters in Equation (4). By including a G protein activation step in the steady state model, we effectively alter the apparent K_G in the equilibrium model [see Equation (7)], possibly causing an agonist inversion. Similarly, by adding a G protein activation step into the model, we effectively change the G protein concentration, opening up another route to agonist inversion.

A second explanation for the different predictions in Figure 3.8 is that the response in the equilibrium model is assumed to be proportional to [R * G] + [LR * G], while in the steady state model the response is assumed to be proportional to [α – GTP]. The reason for this difference is that the steady state model follows the signal transduction pathway farther into the cell, thus we would expect the [α – GTP] concentration to better represent the cell's response than [R * G] + [LR * G]. Although [α – GTP] is clearly related to [R * G] + [LR * G], the two measures of response are coupled nonlinearly because [α – GTP] can also be irreversibly depleted to [α – GDP]. The result of this nonlinear coupling is that [α – GTP] should not identically track [R * G] + [LR * G]; or said more generally, downstream responses generally will not be proportional to the concentration of upstream signaling molecules.

Thus the development of a simple steady state model for G protein activation (Figure 3.5b) allows us to demonstrate the limitations of the equilibrium model as well as improve our ability to model the process. We have used the simplified

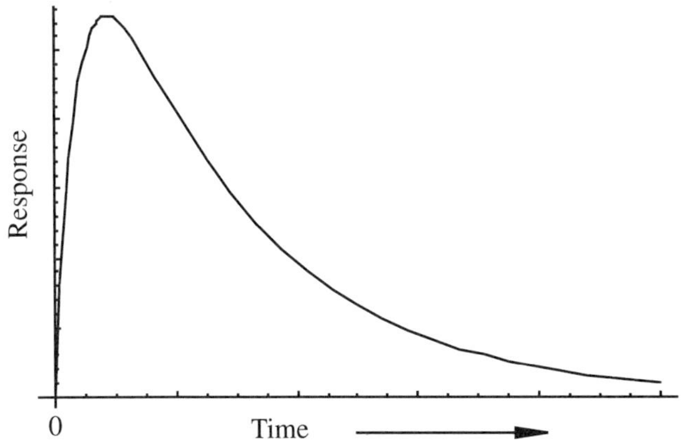

FIGURE 3.9 Desensitization in GPCRs. Persistent agonist stimulation initially causes the response (proportional to the concentration of ligand-bound active receptor [LR] in this case) to increase, however with time, proteins inside the cell deactivate the signaling receptors causing the signal to drop, thereby desensitizing the receptor.

equilibrium criterion of Equation (4) to qualitatively explain some of the behavior in the steady state model. The ability to use a simpler model as a guide in the analysis of more complex behavior is a common theme in modeling and represents a very powerful tool for both the theorist and experimentalist.

3.4 UNSTEADY STATE MODELS

In reality, most biological systems are constantly changing with time, and as such never reach a true steady state. The cells in our body are constantly being bombarded with new information, which causes the cell to change its behavior dynamically. The process of signal transduction is an excellent example of this, as it includes both activation and desensitization dynamics (Figure 3.9). Assuming complete desensitization can occur, the steady state solution would be simply that of no cellular response. Thus analyzing a system with desensitization requires that an unsteady state model be used.

Returning first to our simple model of receptor-ligand binding, we can now add in the simple desensitization step shown in Figure 3.4c. The dynamics of this process are described by the two coupled ordinary differential equations:

$$\begin{aligned}\frac{d[R]}{dt} &= k_{-1}[LR] - k_1[L][R] \\ \frac{d[LR]}{dt} &= k_1[L][R] - k_{-1}[LR] - k_{ds}[LR]\end{aligned} \tag{8}$$

with initial (time = 0) conditions [R] = Rtot and [LR] = 0. Because receptor desensitization is essentially the depletion of free receptors, [R] must be allowed to

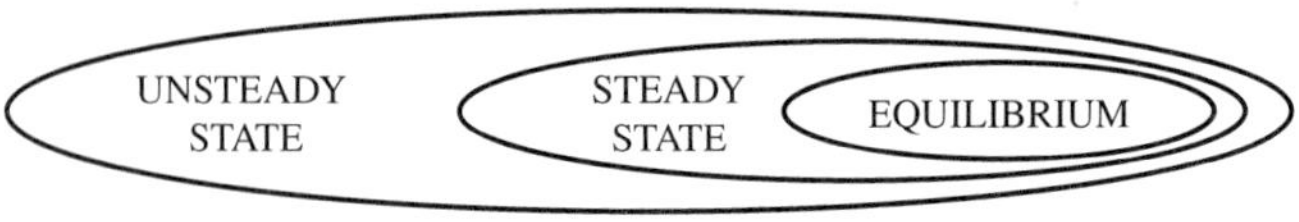

FIGURE 3.10 The relationship between equilibrium, steady state, and unsteady state models.

change with time — therefore species conservation on the receptor must be included. Fortunately, the form of the differential equations in Equation (8) automatically accounts for this, so no further terms must be added. Assuming no significant ligand depletion ([L] is constant), these differential equations are linear and as such are amenable to an analytical solution for [R](t) and [LR](t). The latter is plotted in Figure 3.9.

Note that the unsteady state model becomes the equilibrium model if k_{ds} is assumed to be zero and time is taken to be infinity. Similarly, if k_{ds} is set to 0 and k_1 to $k_{1'} + k_{trans}$, we can recover the steady state model in the limit of infinite time. Thus the unsteady state model is a more general view of the system, and the steady state and equilibrium models are simply subsets of the unsteady state model (see Figure 3.10).

To find the parameters for an unsteady state model, we need to gather time varying (kinetic) data. For example, we and others have generated high quality kinetic data for the binding of ligands to the N-formyl peptide receptor on human neutrophils.[25-28] An advantage of this system is the ability to use fluorescently labeled ligands and to monitor binding with flow cytometry.

We have modeled desensitization and activation in GPCRs using a simplified version of the model shown in Figure 3.5c.[29] This model is unique in that it includes a desensitization step* and as such represents an inherently unsteady state process. From a pharmacological point of view, we are likely most interested in the transient response that occurs prior to significant desensitization, and this can only be simulated using an unsteady state model.

This model was constructed in an attempt to understand the mechanistic basis for ligand efficacy. We and others have hypothesized that ligands may differ in efficacy because they have different values of the conformational selectivity parameter, α, and/or different values of the desensitization rate constant k_{ds}.[24] The ligand-specific parameter α was discussed earlier. The desensitization rate constant may also be influenced by the identity of the ligand bound to the receptor. For example, the receptor may be held in slightly different conformations by different ligands, or the differing dissociation rate constants of ligands may allow for different spatial organization of membrane species relevant to desensitization.[30-32] Ligands with high values of α or small values of k_{ds} are predicted to have greater efficacy than ligands with small values of α or large values of k_{ds}. However, because only active (R*) receptors are believed to desensitize, the effects of k_{ds} cannot be analyzed indepen-

* Only desensitization of the LR* form of the receptor is shown. For constitutively active systems, desensitization via R* should also be included, but was omitted here for simplicity.

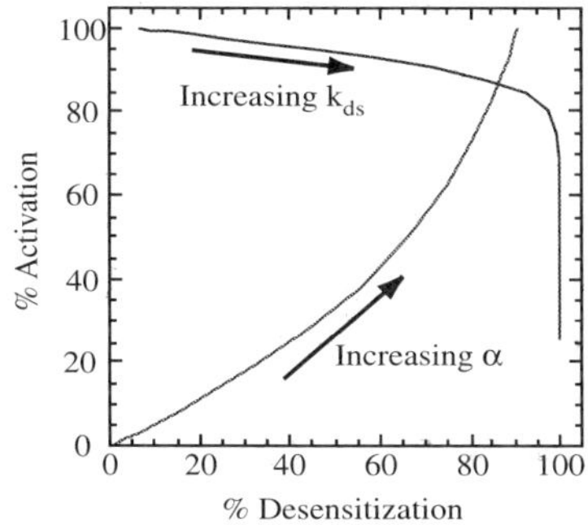

FIGURE 3.11 The simultaneous effects of changing the conformational selectivity factor α and the desensitization rate constant, k_{ds}, on receptor activation and desensitization.[29]

dently of α. Our model can be used to understand the synergy between these two parameters in determining the activation/desensitization profile of a given drug.

The predictions of the GPCR unsteady state model are shown in Figure 3.11. As expected, both α and k_{ds} affect receptor activation and desensitization. In all cases, increasing α causes both more activation and more desensitization, whereas increasing k_{ds} causes less activation and more desensitization.

In order to use these predictions to understand ligand efficacy, we compared the results of Figure 3.11 with data from GPCR systems in which both activation and desensitization have been measured. Literature data on the β_2-adrenergic receptor,[33] μ-opioid receptor,[34] and the dopamine D_1 receptor[35,36] all show the trend of activation and desensitization increasing in parallel, suggesting that the primary differences between ligands in these particular systems is the value of the conformational selectivity α of the ligand. Our own data with the N-formyl peptide receptor system also support this conclusion.[29] Thus the interpretation of data with this unsteady state model has suggested a mechanism behind ligand efficacy in these systems.

It is interesting to note that although the ability of a receptor to transduce a response depends on its "state" (e.g., active, desensitized) and this state may be influenced by the identity of the ligand that is bound, isolating these individual states has proven to be exceedingly difficult. However, combining experimental data for response activation and desensitization with a mathematical modeling approach can provide insight into these receptor states.

Results such as these will ultimately give us the ability to quantitatively specify the properties we want in a drug. For example, in some applications we may want the cell to desensitize very slowly to a highly active drug. How do we find a drug like this? The modeling results guide us to look for drugs with a high value of α and low value of k_{ds}. By comparing the relative α and k_{ds} values of other drugs it might be possible to infer what chemical motifs are required to make a drug with a particular set of properties. Similarly, it may be possible to generate customized high-throughput screening assays that would preferentially detect drugs with specific values of α and k_{ds}. This ability to rationally design and screen for new drugs would be of great assistance to the pharmaceutical industry.

3.5 LOOKING AHEAD

Where will pharmacological modeling take us in the future? Our ultimate goal is to develop predictive pharmacological models that bridge molecular biology (e.g., the details of the receptor-ligand interaction) and cellular biology (e.g., the ultimate cellular response generated by ligand binding). Predictive models would give us a clear view of how a drug affects a cell, allowing us to relate individual drug parameters such as the conformation selectivity parameter α to cellular behavior. Such models would be useful in drug screening as well as for testing complicated drug therapies.

In this chapter, we have discussed three ways to model the role of receptor/ligand/G protein interactions in determining how a cell will respond to a drug, but these models can make only limited predictions because they are based in part on assumptions that make the problem simpler to solve. For example, as discussed in this chapter, sometimes we assume that a particular molecular species is present at a constant concentration (i.e., no depletion/no species conservation) in complex models in order to find a simple solution. One of the clearest assumptions made in current GPCR models is that we can get away with modeling only the receptor, ligand, and G protein in isolation from the rest of the cell. To some extent this approximation is reasonable: the ligand initiates the signaling cascade, and therefore events that take place soon after ligand binding are the events that are most influenced by the specific properties of that ligand. However, the activity of many other proteins inside the cell can have a profound effect on how the receptor's signal is interpreted. In the case of GPCRs, it is known that the receptor is phosphorylated by a number of receptor kinases[37,38] and then binds to adapter proteins, which may have further signaling ability.[39,40] Similarly, proteins downstream of the G protein may also be regulated (by this or other pathways) thereby changing how a particular cell will respond to a drug. For example, downstream proteins may ultimately provide the feedback that regulates the G protein concentration, possibly causing agonist inversion. By including the effects of these other proteins in future models, we will gain a more realistic picture of how the cell interprets its world.

The more molecular species included in our picture of signaling, the more apparent the value of modeling. By including the roles of many different proteins, we generate a complicated reaction network full of feedback and feed forward control loops — a system that is nearly impossible to understand without modeling. This system is not amenable to a reductionist solution, for it is the *interaction* of these many different proteins that give rise to the cellular response. Therefore, pharmacological models *must* move toward analyzing the collective behavior of the many parts in this complex system.

A second assumption we often make is that membrane-associated proteins (e.g., receptors and G proteins) are evenly distributed over the membrane. However, this is unlikely to be true. Most proteins throughout the cell are trafficked to specific locations on the membrane, where they often cluster into groups. By clustering proteins, the cell gains the ability to accelerate reactions well beyond what could be achieved under well-mixed conditions, and to control which reactions take place.[41,42]

For GPCRs, spatial organization plays key roles in both receptor number regulation and receptor cross talk. GPCRs are often removed from the cell surface by clustering into clathrin coated pits, and then internalized via endocytosis. In this case, the spatial organization of the receptors on the cell determines whether the receptor stays on the surface or is internalized. There is also evidence that clustered receptors may be able to activate some tyrosine kinase receptors[9,43,44] in a form of receptor cross talk. This cross talk between different receptor types may be essential to transducing some signals, and as such is pharmacologically important

A more subtle spatial effect is a phenomenon called "switching".[30-32,45-47] When a GPCR is activated by a ligand, the receptor tends to activate all nearby G proteins. If the ligand remains bound to the receptor, then eventually all of the local G proteins will be consumed, leaving the ligand-bound receptor with nothing more to activate. In contrast, if the ligand quickly dissociates from the receptor and binds to a new receptor, then the G proteins around a particular receptor are less likely to be depleted and the signal can continue to be propagated efficiently. Thus a highly active ligand that dissociates slowly may have less of an effect than a less active ligand that dissociates quickly, for a given number of ligand-bound receptors. Only by including spatial effects in our models can we simulate these phenomena.

To make these future models possible will require both more quantitative biological data and faster computers. Although qualitative data are often sufficient to describe the *framework* of a model, it is the quantitative data that describe the *details* of that model. Therefore, quantitative biological data such as *absolute* rates of phosphorylation, binding, dissociation, and conformational change are of key importance if we hope to generate predictive pharmacological models. Unfortunately, biology has historically focused primarily on gathering qualitative data and has only recently begun to report quantitative* results. However, technical advances in biological research should allow for the generation of more quantitative biological data in the future, opening up the possibility of developing truly predictive models.

From a computational standpoint, pharmacological models could be some of the most laborious to simulate. An accurate pharmacological model may need to account for the location and state of hundreds of different kinds of proteins, each replicated tens of thousands of times throughout the cell. This is akin to simulating the dynamic motion of the thousands of parts of an airplane — a simulation that is indeed now possible. Thus, although the computational scale of the problem is daunting, it is most likely only a matter of time until predictive pharmacological models become a standard tool for drug development.

* Here we define quantitative data as absolute, numerical values. Thus when measuring receptor levels, quantitative data would provide the absolute number of receptors (i.e., 32,000 receptors/cell), whereas qualitative data could be relative values, such as the receptor number increased 34%. Although both results are numerical, the absolute receptor count is useful for modeling, while the relative value's main use is to illustrate a qualitative point.

REFERENCES

1. Collins, F. S. and Jegalian, K. G., Deciphering the code of life, *Scientific American,* 281, 86–91, 1999.
2. Lesch, K. P. and Manji, H. K., Signal-transducing G proteins and antidepressant drugs: evidence for modulation of a subunit gene expression rat brain, *Biol. Psychiatry,* 32, 549–79, 1992.
3. Rousch, W., Regulating G protein signaling [news], *Science,* 271, 1056–8, 1996.
4. Neer, E. J., Heterotrimeric G proteins: organizers of transmembrane signals, *Cell,* 80, 249–57, 1995.
5. Berman, D. M. and Gilman, A. G., Mammalian RGS proteins: barbarians at the gate, *J. Biol. Chem.,* 273, 1269–72, 1998.
6. Nestler, E. J. and Aghajanian, G. K., Molecular and cellular basis of addition, *Science,* 278, 58–63, 1997.
7. Chuang, T. T., Iacovelli, L., Sallese, M., and De Blasi, A., G protein-coupled receptors: heterologous regulation of homologous desensitization and its implications, *Trends Pharmacol. Sci.,* 17, 416–21, 1996.
8. Dohlman, H. G. and Thorner, J., RGS proteins and signaling by heterotrimeric G proteins, *J. Biol. Chem.,* 272, 3871–4, 1997.
9. Hall, R. A., Premont, R. T., and Lefkowitz, R. J., Heptahelical receptor signaling: beyond G protein paradigm, *J. Cell Biol.,* 145, 927–32, 1999.
10. Lauffenburger, D. A. and Linderman, J. J., *Receptors: Models for Binding, Trafficking, and Signaling,* Oxford University Press, New York, 1993.
11. Samama, P., Cotecchia, S., Costa, T., and Lefkowitz, R. J., A mutation-induced activated state of the beta 2-adrenergic receptor. Extending the ternary complex model, *J. Biol. Chem.,* 268, 4625–36, 1993.
12. Lefkowitz, R. J., Cotecchia, S., Samama, P., and Costa, T., Constitutive activity of receptors coupled to guanine nucleotide regulatory proteins, *Trends Pharmacol. Sci.,* 14, 303–7, 1993.
13. Kenakin, T. P., Receptor conformational induction versus selection: all part of the same energy landscape, *Trends Pharm. Sci.,* 17, 190–1, 1996.
14. Kenakin, T., The classification of seven transmembrane receptors in recombinant expression systems, *Pharmacol. Rev.,* 48, 413–63, 1996.
15. Weiss, J. M., Morgan, P. H., Lutz, M. W., and Kenakin, T. P., The cubic ternary complex receptor-occupancy model. I. Model Description, *J. Theor. Biol.,* 178, 151–67, 1996.
16. Weiss, J. M., Morgan, P. H., Lutz, M. W., and Kenakin, T. P., The cubic ternary complex receptor-occupancy model. III. Resurrecting efficacy, *J. Theor. Biol.,* 181, 381–97, 1996.
17. Kenakin, T., Pharmacological proteus?, *Trends Pharmacol. Sci.,* 16, 256–8, 1995.
18. Kenakin, T., Protean agonists. Keys to receptor active states?, *Ann. NY Acad. Sci.,* 812, 116–25, 1997.
19. Yatsui, K., Becker, E. L., and Sha'afi, R. I., Lipopolysaccharide and serum cause the translocation of G protein to the membrane and prime neutrophils via CD14, *Biochem. Biophys. Res. Comm.,* 183, 1280–6, 1992.
20. Durstin, M., McColl, S. R., Gomez-Cambronero, J., Naccache, P. H., and Sha'afi, R. I., Up-regulation of the amount of Gia2 associated with the plasma membrane in human neutrophils stimulated by granulocyte-macrophage colony stimulating factor, *Biochem. J.,* 292, 183–7, 1993.

21. Chidiac, P., Nouet, S., and Bouvier, M., Agonist-induced modulation of inverse agonist efficacy at the beta 2-adrenergic receptor, *Mol. Pharmacol.,* 50, 662–9, 1996.
22. Adams, J. A., Omann, G. M., and Linderman, J. J., A mathematical model for ligand/receptor/G protein dynamics and actin polymerization in human neutrophils, *J. Theor. Biol.,* 193, 543-–60, 1998.
23. Shea, L. D., Neubig, R. R., and Linderman, J. J., Timing is everything: the role of kinetics in G protein activation, submitted, 2000.
24. Linderman, J. J., Kinetic approaches to understanding ligand efficacy, in *Handbook of Experimental Pharmacology,* T. P. Kenakin and J. A. Angus, Eds, Springer-Verlag, 2000, in press.
25. Hoffman, J. F., Linderman, J. J., and Omann, G. M., Receptor up-regulation, internalization, and interconverting receptor states. Critical components of a quantitative description of N-formyl peptide-receptor dynamics in the neutrophil, *J. Biol. Chem.,* 271, 18394–404, 1996.
26. Hoffman, J. F., Keil, M. L., Riccobene, T. A., Omann, G. M., and Linderman, J. J., Interconverting receptor states at 4 degrees C for the neutrophil N-formyl peptide receptor, *Biochemistry,* 35, 13047–55, 1996.
27. Fay, S. P., Posner, R. G., Swann, W. N., and Sklar, L. A., Real-time analysis of the assembly of ligand, receptor, and G protein by quantitative fluorescence flow cytometry, *Biochem.,* 30, 5066–75, 1991.
28. Neubig, R. R. and Sklar, L. A., Subsecond modulation of formyl peptide-linked guanine nucleotide-binding proteins by quanosine 5′-O-(3-thio)triphosphate in permeabilized neutrophils, *Molec. Pharm.,* 43, 734–40, 1993.
29. Riccobene, T. A., Omann, G. M., and Linderman, J. J., Modeling activation and desensitization of G protein coupled receptors provides insight into ligand efficacy, *J. Theor. Biol.,* 200, 207–22, 1999.
30. Shea, L. D. and Linderman, J. J., Compartmentalization of receptors and enzymes affects activation for a collision coupling mechanism, *J. Theor. Biol.,* 191, 249–58, 1998.
31. Shea, L. and Linderman, J. J., Mechanistic model of G protein signal transduction. Determinants of efficacy and effect of precoupled receptors, *Biochem. Pharmacol.,* 53, 519–30, 1997.
32. Shea, L. D., Omann, G. M., and Linderman, J. J., Calculation of diffusion-limited kinetics for the reactions in collision coupling and receptor cross-linking, *Biophys. J.,* 73, 2949–59, 1997.
33. Benovic, J. L., Staniszewski, C., Mayor, J., F., Caron, M. G., and Lefkowitz, R. J., Beta-adrenergic receptor kinase. Activity of partial agonists for stimulation of adenylate cyclase correlates with ability to promote receptor phosphorylation, *J. Biol. Chem.,* 263, 3893–7, 1988.
34. Yu, Y., Zhang, L., Yin, S., Sun, H., Uhl, G. R., and Wang, J. B., μ opioid receptor phosphorylation, desensitization, and ligand efficacy, *J. Biol. Chem.,* 272, 28869–74, 1997.
35. Balmforth, A. J., Warburton, P., and Ball, S. G., Homologous desensitization of the D1 dopamine receptor, *J. Neurochem.,* 55, 2111–16, 1990.
36. Barton, A. C. and Sibley, D. R., Agonist-induced desensitization of D1-dopamine receptors linked to adenyl cyclase activity in cultured NS20Y neuroblastoma cells, *Mol. Pharm.,* 38, 531–41, 1990.
37. Lefkowitz, R. J., G protein-coupled receptors. III. New roles for receptor kinases and beta-arrestins in receptor signaling and desensitization, *J. Biol. Chem.,* 273, 18677–80, 1998.

38. Krupnick, J. G. and Benovic, J. L., The role of receptor kinases and arrestins in G protein-coupled receptor regulation, *Ann. Rev. Pharmacol. Toxicol.,* 38, 289–319, 1998.
39. Lefkowitz, R. J., Inglese, J., Koch, W. J., Pitcher, J., Attramadal, H., and Caron, M. G., G protein-coupled receptors: regulatory role of receptor kinases and arrestin proteins, *Cold Spring Harb. Symp. Quant. Biol.,* 57, 127–33, 1992.
40. Zuker, C. S. and Ranganathan, R., The path to specificity [comment], *Science,* 283, 650–1, 1999.
41. Bray, D., Signaling complexes: biophysical constraints on intracellular communication, *Ann. Rev. Biophys. Biomol. Struct.,* 27, 59–75, 1998.
42. Allison, S. A., Northrup, S. H., and McCammon, J. A., Stimulation of biomolecular diffusion and complex formation, *Biophys. J.,* 49, 167–75, 1986.
43. Luttrell, L. M., Daaka, Y., and Lefkowitz, R. J., Regulation of tyrosine kinase cascades by G protein-coupled receptors, *Curr. Opin. Cell. Biol.,* 11, 177–83, 1999.
44. Lin, F. T., Miller, W. E., Luttrell, L. M., and Lefkowitz, R. J., Feedback regulation of beta-arrestin 1 function by extracellular signal — regulated kinases, *J. Biol. Chem.,* 274, 15971–4, 1999.
45. Stickle, D. and Barber, R., Evidence for the role of epinephrine binding frequency in activation of adenylate cyclase, *Mol. Pharm.,* 36, 437–45, 1989.
46. Mahama, P. A. and Linderman, J. J., A Monte Carlo study of the dynamics of G protein activation, *Biophys. J.,* 6, 1345–57, 1994.
47. Mahama, P. A. and Linderman, J. J., Monte Carlo simulations of membrane signal transduction events: effect of receptor blockers on G protein activation, *Ann. Biomed. Eng.,* 23, 299–307, 1995.

4 Intramolecular Dynamics and Ligand-Induced Conformational Changes: A Stochastic Model of Receptor Action

H. Ongun Onaran and Tommaso Costa

CONTENTS

4.1 INTRODUCTION

Quantitative models of protein action represent an important means of biochemical research. They provide not only valuable insight into fundamental questions of biological science, such as enzyme kinetics and mechanisms, and control of channel activity and receptor action, but also foster progress in areas of medical research that have a direct impact on human health, such as the procedure of drug discovery and rational pharmacological therapy.

Despite the variety of biological activities in which they are involved, all proteins share a common process that underlies their functioning: the ability to change chemical reactivity to their environment in response to interactions with ligands. Proteins bind their ligands with remarkable specificity and selectivity, which implies

0-8493-0100-9/01/$0.00+$.50

a very high degree of structural complementarity in the atomic configurations of the interacting sites. But they also act in response to such interactions with rapid changes of functional state, which requires dynamic flexibility of their molecular architectures. This paradoxical coexistence of stringency and degeneracy in protein structures is the core problem in quantitative models of protein activity.

Binding interactions between protein ligands can effectively be described in the abstract layer of macroscopic thermodynamics (e.g., mass action law), even though they actually result from multiple and heterogeneous bonds among numerous residues of the reacting molecules. But the perturbation of intramolecular energy that triggers the modification of biological reactivity on binding needs statistical mechanics theory to be defined precisely. Thus "binding" and the "molecular consequences of binding" often appear as separate mechanisms, although they result from the same forces and represent an inseparable physical process.

Modern biophysicists define proteins — and biopolymers in general — as complex systems, since these molecules do not exist in a well-defined energy minimum, but in a very large number of conformationally similar substates of approximately equal energy.[1,2] Future progress in the theory and analysis of complex systems in physics will likely bring dramatic improvements in our understanding of biological macromolecules. For the time being, however, the exact meaning of the term "conformational change" remains largely obscure, despite its widespread use in biology.

Although thermodynamics does not allow one to specify how intramolecular interactions make up the chemical potential of a protein, it can predict their effect on the free energy changes driving the reactions of the protein with its interacting partners. It is thus possible, at least in principle, to define thermodynamic potentials that measure what portion of free energy change is due to intramolecular contributions. A landmark of such an approach is the binding potential of Wyman, which stands behind the concept of allosteric linkage and all models that are based upon it.[3,4] Further introduction of "allosteric" states into the models that deal with allosteric linkage[5,6] provides an elegant simplification of protein behaviour and can be applied to the most complex problems, including folding.[7]

However, the widespread use of protein engineering experiments and the large collection of accumulating data focus the attention of research more and more toward the role that individual residues of the protein sequence play in their overall biological activity. The fundamental question is how to relate structure and motion of protein molecules to their function. Allosteric models based on the definition of allosteric states imply a deterministic correlation between configurational and functional states of the protein, but in a complex system, the relation between microscopic constituents and its macroscopic behaviour is stochastic. A stochastic model, therefore, may provide a more realistic picture of the relation between structure/dynamics changes in proteins and their function.

In this paper we describe a model of protein action that is based on a probabilistic view of the conformational space of a macromolecule. We show that the model can be either related to macroscopic allosteric behaviour through equilibrium analysis, or can be used to analyse the kinetics of state transitions in proteins through non-stationary formulation. Although the ideas proposed in the present paper apply to all functional proteins (such as enzymes, ion channels, receptors, etc.), we give particular emphasis to G protein-coupled receptors and refer to experimental observations made for this important class of proteins.

4.2 RECEPTOR MODELS AND CHEMICAL THERMODYNAMICS OF CONFORMATIONAL CHANGES

As we mentioned above, any model that has been intended to explain receptor action deals with changes in reactivity of protein molecules to their environment upon ligation. In a very broad sense, this can be seen as ligand-mediated modulation of protein functions. In the specific case of G protein-coupled receptors (GPCR), which, upon occupation by an agonist, interact with G proteins and desensitizing GPCR kinases, the change of reactivity is observed as modified chemical affinity for such signal transduction partners.[8-14] The ability of a ligand to modify the reactivity state of a receptor has been termed molecular efficacy, and can be seen — at least in the imagination — as an entity separate from the ability to bind the receptor itself. Thus, affinity and efficacy become the fundamental qualifiers of ligands acting on a signal transduction system. These are the main concepts behind all the equilibrium models of receptor action proposed so far.[5,15-32]

We can roughly classify these models in two categories without entering into the details of the phenomena they explain:

- Ternary complex model
- Allosteric models

The ternary complex model poses the question "what difference between empty and ligated receptor (i.e., R and HR) accounts for an agonist-induced change of signaling status of the receptor?" The answer is that — for any unspecified mechanism whatsoever — R can have a different affinity for G protein (G) compared to HR by a factor of α (a positive real number). The formulation is based on the principles of chemical thermodynamics, which require reciprocity in the reactivities of the receptor for the two ligands H and G, i.e., the affinities of R and RG for H differ by the same factor, α, as those of R and HR for G (see Figure 4.1).[28,33-37] It is implicit in the model that if a difference exists ($\alpha \neq 1$), there must be a binding-induced conformational change in the receptor that explains why it is so.

Allosteric models add the concept of ligand-mediated conformational change in the receptor as an explicit mechanism: the receptor is declared to exist in different "states," the proportion of which can be changed by ligand binding.[3-6,24,38] Thus, allosteric models provide a slightly different answer to the same fundamental question: i.e., R and HR react differenly because the abundance of different "allosteric states" in the empty receptor form differs from that of the ligated form. In such models, allosteric states of the protein are considered in equilibrium with each other and assumed to possess preassigned properties (e.g., "bindability" for G protein, ability to conduct ions for ion channels, catalytic competence for enzymes, etc.). Ligands are there to modify the equilibrium among allosteric states (not their nature!)

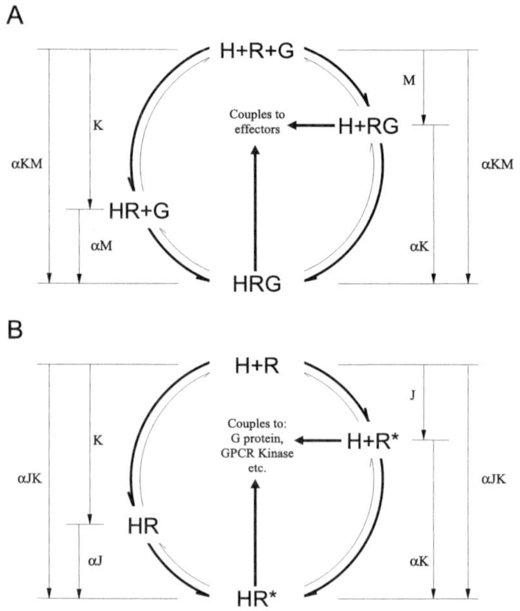

FIGURE 4.1 A. Thermodynamic cycle of the ternary complex model of G protein-coupled receptor activation. R, represents receptor, which binds ligand (H) and G protein (G). Equilibrium affinity constants are shown next to reversible reaction arrows. Formation of the ternary complex (HRG) is achieved by two different paths: H+R+G=HR+G=HRG or H+R+G=H+RG=HRG. These two paths are not independent in the sense that they imply the same change of standard free energy of formation of the ternary complex from the free species. Therefore, the overall reaction affinity constants for the two paths must be the same (i.e., αKM). Note that affinity constants are multiplicative, whereas free energy changes are additive along the paths. B. Thermodynamic cycle of two-state allosteric model. The receptor is assumed to exist in two allosteric states (R and R*) which are identified by their "functional abilities." R* is assumed to be the one that can bind G protein or GPCR kinase as indicated in the picture. Ligand (H) is assumed to modify the equilibrium between R and R*. The thermodynamic schematization and parametrization of the whole cycle is as in (A).

and thus can alter the observable functioning of the protein. Again, the formulations are based on the principle of chemical thermodynamics,[4] which means that the two classes of models are based on identical theoretical background and indeed represent two faces of the same coin (see Figure 4.1).

It is important to understand the theoretical approach behind such models. Perturbations of intramolecular energy distributions (i.e., what causes a conformational change) in proteins and its relation to binding affinity can be expressed in classical thermodynamics as the modifications of the expected free energy changes that would take place if the protein existed in a fixed conformation of its atomic structure. The allosteric models, therefore, are a strategy to express in simple mass action law-based terms, a phenomenon (conformational effect on binding affinity and vice-versa) that would otherwise need the partition functions of statistical mechanics to be rigorously addressed. As a consequence, the allosteric states that are defined by these models are not true physical states, but rather functional states, and the allosteric constants thus derived are average macroscopic quantities related to the observed biological function more than to the physical properties of the protein.

Therefore, allosteric models do not tell why empty and free receptors behave differently, but only provide a rule to compute the extent and the direction of the difference. Since allosteric states themselves are not "pure" states in the strict physical sense, problems arise when we try to associate the functional behaviour of the protein to some detailed physical properties measured, for example, by direct spectroscopic methods.

4.3 MULTISTATE PROBABILISTIC MODEL OF RECEPTOR ACTION

Thus, we are back to the original question: what is the difference between R* and HR* (Figure 4.1B)? The root of the problem is that a receptor can be viewed as a dynamic object in persistent exploration of a very large number of energetically similar conformations, and there is no way to tell whether what we call "function" is the attribution of some specific configurations or the result of how the receptor travels across its conformational space. To reflect such uncertainty in the relation between physics and function, we can directly start with an unspecified number of states that a protein can theoretically assume, without making any preassignment of functional properties to individual states.[1,2] Then, "function" arises as a macroscopic result of the distribution of these "microscopic" states under different conditions. This concept is the main idea behind the multistate probabilistic model, and its formulation is based on state probabilities rather than macroscopic reaction theory. Otherwise, the mathematical formulation of the model is exactly the same as that of the allosteric models. In the following few paragraphs we briefly summarize the formulation, and illustrate the basic equilibrium properties of the probabilistic model. We give particular reference to the case of GPCR, where there is a bivalent protein (receptor) that binds ligands and G proteins. The reader can refer to References 30 and 39 for a more detailed discussion.

Let us call the set of *all* theoretically possible states of a receptor the "conformational space" (S) of the protein, and assume that each member s_i of S is occupied with a time-independent probability (p_i) in thermodynamic equilibrium. Then, the collection p of p_i defines a time-independent probability distribution over S of the receptor protein (R). In a population of such receptors, these probabilities are given as $[s_i]/[R]$ (where brackets are read as the number or concentration of the corresponding species). Given the probability distribution p, we can further define a set of constants (j_i) that govern the equilibrium transitions between an arbitrary reference state (s_0) and all the others (s_i, $i \neq 0$): $j_i = [s_i]/[s_0] = p_i/p_0$, with $j_0 = 1$. Binding of a ligand X (either H or G) to the receptor can change the distribution of these states. Therefore, the (conditional) probability distribution of X-bound receptor (${}^xp$) over S should be defined separately as ${}^xp_i = [{}^xs_i]/[XR]$ and ${}^xj_i = [{}^xs_i]/[{}^xs_0] = {}^xp_i/{}^xp_0$, (with ${}^xj_0 = 1$), where left-hand superscripts indicate the presence of bound ligand. Finally, by defining the effect that ligand binding imparts on the equilibrium between s_0 and s_i as ${}^xb_i = {}^xj_i / j_i$, we can give the macroscopic properties of the receptor in terms of the probabilities defined above. By macroscopic properties we mean the macroscopic affinities of formation of ligand-receptor (HR) or receptor-G protein (RG) complex ($\mathbf{K}_h$ or $\mathbf{K}_g$, respectively), and the effect of H and G on the affinities of each other when they are simultaneously bound to the receptor (i.e., the factor α defined above):

$$\alpha = (\Omega\Omega_{hg})(\Omega_h\Omega_g)^{-1} \tag{1}$$

$$\mathbf{K}_h = {}^hk_0\Omega_h(\Omega)^{-1} \tag{2a}$$

$$\mathbf{K}_g = {}^gk_0\Omega_g(\Omega)^{-1} \tag{2b}$$

The *virtual* constants for the formation of ligand-s_0 complexes are defined as ${}^hk_0 = [{}^hs_0]/([H][s_0])$ and ${}^gk_0 = [{}^gs_0]/([G][s_0])$ for ligands H and G respectively. The quantities Ω's are defined as follows:

$$\Omega = 1 + \sum_i j_i,\ \Omega_h = 1 + \Omega\sum_i {}^hb_ip_i,\ \Omega_g = 1 + \Omega\sum_i {}^gb_ip_i,\ \Omega_{hg} = 1 + \Omega\sum_i {}^hb_i{}^gb_ip_i$$

where summations run over the conformational space, save the reference state (i.e., $i \neq 0$). The superscript h or g is used instead of X to denote particular ligands (i.e., H or G). The constant α can be given in a more compact (and useful) form as:

$$\alpha = \sum_S \frac{{}^hp_i\,{}^gp_i}{p_i} \tag{3}$$

where S indicates that the summation runs over the entire conformational space (i.e., no reference state is assumed).

We can summarize the basic concepts and implications of this model as follows:

1. A protein (the receptor in our case), in its naturally folded form, can be considered as a random machine that constantly explores a conformational space consisting of a finite number of states. The equilibrium description of the protein can be given in terms of the probabilities of occupation of such states in thermal equilibrium, and thus defines a probability distribution over the conformational space. Ligand binding to the protein can change this distribution.
2. In a bivalent protein, the probability distributions of the free and different bound states (i.e., the empty, ligand1-bound, ligand2-bound forms of the protein), uniquely determine the interaction of two of the ligands on the protein when they are simultaneously bound [Equation (1) or (3)]. In other words, the origin of the allosteric interactions of two ligands binding on separate sites of a protein can be seen as the difference between the probability distributions that the protein assumes at different ligation states.

3. For a given set of α, p_i, and ${}^g p_i$, Equation (3) has more than one solution for the probability distribution ${}^h p_i$. This implies that for a given receptor-G protein pair (conferred by given p_i and ${}^g p_i$), there may be ligands with different physical effects on the receptor (which are identified by different ${}^h p_i$), possessing the same efficacy, α. Thus, the coupling α may degenerate at the microscopic level.
4. The macroscopic affinity for the formation of ligand-protein complex results from two contributions. One represents the random movements of the protein across the conformational space (hidden in the Ωs of Equations (2a) and (2b) as a function of probability distributions at different ligation states); another from the (virtual) interaction of ligand with a particular state of the conformational space [showing as ${}^x k_0$ in Equations (2a) and (2b)]. In contrast, the macroscopic coupling α results only from the random movements of the protein [see Equation (3)].
5. Thus, any physical intervention that may affect these probability distributions (such as engineered mutations) is expected to affect the affinities of ligands *and* their interactions on the protein in a complicated but correlated way [compare Equation (1) and (2)]. This may be particularly important when attempting to separate and map different aspects of the protein (e.g., ligand binding properties of the protein and ligand-induced changes in protein functions) onto particular sites in the protein by mutational studies. While allosteric models expect perfect correlations, the probabilistic model predicts that such correlations will be blurred or distorted by the complicated relation between internal movement of the protein and functional outcome.

We recently used these concepts to explain the effects of particular point mutations in the α_{1b} adrenoceptor (α_{1b}-AR) on macroscopic ligand affinities and ligand efficacies, which could hardly be explained by the allosteric activation models proposed so far for GPCR.[39] In the following section, we give a dynamic formulation for the multistate-probabilistic model, by which time-dependent properties of the model can be analyzed.

4.4 DYNAMIC FORMULATION OF THE MULTISTATE MODEL

In the previous section, we showed how the equilibrium properties of the multistate model can be described in terms of probability distributions of microscopic states. Here, we analyze the dynamic properties of such a system. We shall attempt to define how the distribution of microstates evolves in time from a given initial distribution to a final target distribution.

Let us denote by $\mathbf{x}(t)$ the stochastic process describing the random shuttling of a macromolecule among the states (s_i) of the conformational space (S). Then, $\mathbf{x}(t)$ is a discrete-state continuous-time process, for we assumed that the conformational space of the protein consists of countable states. The time history of such a process can be schematized as in Figure 4.2.

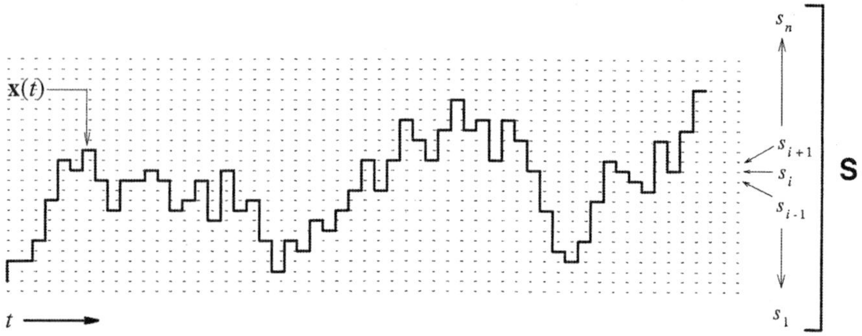

FIGURE 4.2 Example of a discrete-state continous time stochastic process. The process, denoted as x(t), randomly shuttles between the discrete states (s_i) that the process can possibly occupy. Transitions between different states are seen as discontinuity points in the process.

This process can be considered such that its past has no influence on its future, if its present is specified. In other words, if the process is in state i at time t, the probability of its leaving or staying in that state at $t + \tau$ ($\tau \geq 0$) does not depend on how the state i was reached. Such an assumption seems reasonable if considered in analogy to the mechanical (classical or quantum) state of a system (protein in the present case), which can be fully specified by a point in the phase space (or in the Hilbert space). The phase point then determines the future states of the system without any reference to how it was reached (see for example reference 40). A stochastic process is called a Markoff process if it possess the properties described above. Thus, the process $\mathbf{x}(t)$ can be considered as a continuous-time, discrete-state Markoff process, and it can be specified by means of two time-dependent probability functions:

i) *state probabilities* $p_i(t) = P\{\mathbf{x}(t) = s_i\}$
ii) *transition probabilities* $\pi_{ij}(t_1,t_2) = P\{\mathbf{x}(t_2) = s_j | \mathbf{x}(t_1) = s_i\}(t_2 > t_1)$

which are the probabilities of the following two events:

{x is in state i at time t} and
{x *is* in state j at time t_2 given that it *was* in state i at time t_1},

respectively. See Figure 4.3 for a schematic illustration of transition probability functions.

These functions have the following important properties to which we shall refer later:

$$\sum_j \pi_{ij}(t_1,t_2) = 1 \tag{4}$$

$$\sum_i p_i(t_1)\pi_{ij}(t_1,t_2) = p_j(t_2) \tag{5}$$

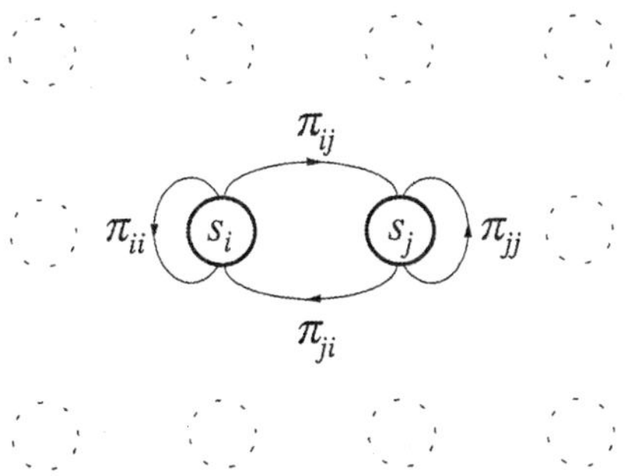

FIGURE 4.3 Schematic representation of transition probabilities for a discrete-state Markoff process. Probability of a transition from state s_i to s_j in a given time interval is denoted as π_{ij} (or vice-versa for the reverse transition) and probability of staying in the same state as π_{ii} or π_{jj}.

$$\pi_{ij}(t_1,t_3) = \sum_r \pi_{ir}(t_1,t_2)\pi_{rj}(t_2,t_3) \qquad t_1 < t_2 < t_3 \tag{6}$$

Equation (4) holds because the process, given that it is in state i at time t_1, can either jump to any other possible state or stay in state i at time t_2 (it has no other choice). Equation (6) is known as the Chapman-Kolmogoroff equation.

In a homogenous Markoff process, the transition probabilities depend on the magnitude of the time interval $\tau = t_2 - t_1$, i.e.,

$$\pi_{ij}(\tau) = P\{\mathbf{x}(t+\tau) = s_j | \mathbf{x}(t) = s_i\} \qquad (\tau \geq 0)$$

and they are given as a matrix $\Pi(\tau)$, called the transition matrix, whose elements are $\pi_{ij}(\tau)$:

$$\Pi(\tau) - \begin{bmatrix} \pi_{11}(\tau) & \cdots & \pi_{1n}(\tau) \\ \cdots & \cdots & \cdots \\ \pi_{n1}(\tau) & \cdots & \pi_{nm}(\tau) \end{bmatrix}$$

If we define $\alpha = t_3 - t_2$, the Chapman-Kolmogoroff equation [Equation (6)] for a discrete-state continuous-time Markoff process then becomes:

$$\pi_{ij}(\tau+\alpha) = \sum_r \pi_{ir}(\tau)\pi_{rj}(\alpha)$$

or in matrix notation:

$$\Pi(\tau+\alpha) = \Pi(\tau)\Pi(\alpha) \qquad \alpha,\ \tau \geq 0 \tag{7}$$

Now the question is how the state probabilities, $\mathbf{P}(t) = [p_1(t), p_2(t), \ldots, p_n(t)]$ evolve in time when the function $\Pi(\tau)$ and initial state probabilities $\mathbf{P}(0) = [p_1(0), p_2(0),$

..., $p_n(0)]$ are given for the entire space, $\mathbf{S}$. The following is a standard treatment of such a problem (see for example Reference 41).

Let us first define the matrix **K** as:

$$\left.\frac{d\Pi(\tau)}{d\tau}\right|_{\tau=0^+} = \mathbf{K} \equiv \begin{bmatrix} k_{11} & \cdots & k_{1n} \\ \cdots & \cdots & \cdots \\ k_{n1} & \cdots & k_{nm} \end{bmatrix}$$

whose elements $k_{ij} = \pi'_{ij}(0^+)$ are the derivatives from the right of the elements $\pi_{ij}(\tau)$ of $\mathbf{\Pi}(\tau)$. The assumption here is that $\pi_{ij}(\tau)$ is differentiable at $\tau = 0^+$, which is so only if the probability that there is one discontinuity point (i.e., a transition occurs) in the interval $[t, t + \tau]$ is in the order of τ. Note that the off-diagonal elements of **K** are simply the rate constants used in chemical kinetics.

Differentiating Equation (7) with respect to α and setting $\alpha \to 0$ we obtain the following differential equation with constant coefficients:

$$\Pi'(\tau) = \Pi(\tau)\mathbf{K} \quad \text{with} \quad \Pi(0) = \mathbf{I} \tag{8}$$

where **I** is the identity matrix, because the process has no option but to stay in whatever state it occupies at time 0, so that we have transition probabilities equal to one at the diagonal and zeros elsewhere:

$$\pi_{ij}(0) = \begin{cases} 1 & i = j \\ 0 & i \neq j \end{cases} \tag{9}$$

From the definition of **K** and from Equation (4) we clearly have:

$$\sum_j k_{ij} = 0$$

and since $k_{ij} \geq 0$ for $i \neq j$ [see Equation (9)] then we must have:

$$-k_{ii} = \sum_j k_{ij} \qquad i \neq j \tag{10}$$

So the diagonal elements of **K** are the negatives of the row-wise-sum of the off-diagonal elements at the corresponding rows.

The solution to the system of n differential equations given in Equation (8) is:

$$\Pi(\tau) = e^{\mathbf{K}\tau} \tag{11}$$

which are the time-dependent transition probabilities in terms of the *usual* rate constants.

As for **P**(t) we have the following equation [from Equation (5)]:

$$\mathbf{P}(t + \tau) = \mathbf{P}(t)\Pi(\tau)$$

and differentiating with respect to τ and setting $\tau \to 0$, as above, we have:

$$\mathbf{P}'(t) = \mathbf{P}(t)\mathbf{K}$$

and again, its general solution is the vector exponential:

$$\mathbf{P}(t) = \mathbf{P}(0)e^{\mathbf{K}t} \tag{12}$$

K is a singular matrix. Thus, at least one of its eigenvalues equals 0. In most cases, especially in ours where many reversible transitions between states are allowed with different rate constants, the n-1 eigenvalues (λ_i) are different. In such cases **K** can be given as:

$$\mathbf{K} = \mathbf{L}\Lambda\mathbf{M}$$

where $\Lambda = \text{diag}(\lambda_1, \lambda_2, \ldots, \lambda_n)$, **L** is a $n \times n$ matrix with columns ($\mathbf{l}_j$) equal to the corresponding eigenvectors of **K** and $\mathbf{M} = \mathbf{L}^{-1}$. Let us also denote by $\mathbf{m}_i$ the rows of **M** and define a set of matrix **A** as:

$$\mathbf{A}_i = \mathbf{l}_i\mathbf{m}_i \qquad i = 1, 2, \ldots, n$$

The so-called spectral expansion of **K** is then given as:

$$\mathbf{K} = \sum_{i=1}^{n} \mathbf{A}_i\lambda_i$$

which, using Equations (11) and (12) can be written in the form:

$$\Pi(\tau) = \sum_{i=1}^{n} \mathbf{A}_i e^{\lambda_i \tau} \tag{13}$$

$$\mathbf{P}(t) = \mathbf{P}(0)\sum_{i=1}^{n} \mathbf{A}_i e^{\lambda_i t} \tag{14}$$

Let us chose the index for the zero eigenvalue as 1 (i.e., $\lambda_1 = 0$). Since $\lambda_i < 0$ for $i > 1$, then all the terms, but the first one, in the summation in Equation (14) vanish as $t \to \infty$. In other words, the first term in Equation (14) gives the equilibrium occupancies of the states for a given **K**:

$$\mathbf{P}(\infty) = \mathbf{P}(0)\mathbf{A}_1 \tag{15}$$

Finally, if the process $\mathbf{x}(t)$ is stationary (i.e., the statistics of $\mathbf{x}(t)$ are identical to that of $\mathbf{x}(t + c)$ for any arbitrary value of c, or equivalently, its statistics are insensitive to a time shift) then $p_i(t) = p_i$ = constant. A system represented by such a process with constant state probabilities is considered at equilibrium. Therefore, the probabilities (p_i) are identical to those that we discussed in the previous section for the equilibrium treatment of the model.

The state probabilities have obvious connections to the concentrations of the states when we consider an assembly of identical and independent protein molecules, i.e., multiplying $\mathbf{P}(t)$ with the total concentration of the protein yields the *expected* concentrations of different states at any given time.

In summary, the random exploration by a protein of its conformational space can be considered as a discrete-state continuous-time Markoff process. Given the matrix **K** and an initial state distribution $\mathbf{P}(0)$ $(p_1, p_2, \ldots, p_i, \ldots)$, the state probabilities can be calculated at any given time by Equation (12) [or by Equation (14)] and the final state distribution by Equation (15). Finally, the distribution of eigenvalues of **K** provides information about the exponential contributions to the overall relaxation of the protein from an initial "state" to a final one (i.e., $\mathbf{P}(0) \rightarrow \mathbf{P}(\infty)$). By "state" here we mean the probability distribution defined over the conformational space of the protein, hence the microstate (s_i) composition of the protein (as in the previous chapter). A similar mathematical formulation has been applied to multistate ion channel kinetics by Colquhoun and Hawks.[42,43]

In the next section, we give some numerical simulations to show how the dynamics of the multistate model can explain slow macroscopic relaxations (order of seconds to minutes) observed in many different proteins despite the expected fast dynamics at the microscopic scale (order of picoseconds), or the asymmetric kinetics of macroscopic state transitions in the β_2-AR observed in fluorescence experiments.

4.5 RELATIONSHIP BETWEEN RATES OF INDIVIDUAL MICROSTATE TRANSITIONS AND OVERALL RELAXATION RATE

Measurements and predictions have shown that spontaneous refolding of proteins occurs in a time-scale range of microseconds to hours.[44-46] However, a ligand-induced reversible "conformational" change in a naturally folded protein may be expected to proceed much faster. Indeed, the typical time-scale used in molecular-dynamic simulations of folded proteins is on the order of picoseconds, which can be considered as the time scale at the microscopic level that seems to be appropriate to simulate many aspects of equilibrium behavior of proteins.[47]

On the other hand, experimental and theoretical studies have shown that ligand-mediated conformational changes in many different proteins, including some membrane receptors, can occur very slowly compared to the picosecond time scale of the microscopic world of the proteins.[48-50] Thus, a tremendous discrepancy becomes apparent between microscopic and macroscopic behavior of the proteins. The multistate model that we discussed above may be of help in understanding the reason for such a discrepancy. Below, we provide some numerical simulations using the multi-state probabilistic model to address this question.

The following strategy was used for the numerical simulations:

1. The matrix **K** of the transition rate constants was constructed randomly (see below).
2. Eigenvalues and corresponding eigenvectors of **K** were calculated and the matrices **L, M,** and **A** were constructed as described above.
3. $\mathbf{P}(\infty)$, which is the equilibrium state distribution imposed by **K**, was calculated using Equation (15) with an arbitrary $\mathbf{P}(0)$.
4. In order to imitate ligand binding, a random perturbation was given to **K,** as $\mathbf{K}+\varepsilon$ (such that $k_{ij} + \varepsilon_{ij} \geq 0$ for all $i \neq j$).
5. The distribution calculated in step 3 [i.e., $\mathbf{P}(\infty)_{old}$] was used as $\mathbf{P}(0)$ and the new distribution $\mathbf{P}(\infty)_{new}$ imposed by the perturbed **K** was calculated by using Equation (15) as above.
6. Entire relaxation from $\mathbf{P}(\infty)_{old} \rightarrow \mathbf{P}(\infty)_{new}$, which simulates ligand-induced change in protein state, was calculated using either Equation (14) or (12).

In some cases, we only analyzed the distributions of eigenvalues (steps 1 and 2 only), without calculating full relaxation processes.

4.5.1 Construction of the Matrix (K) of Rate Constants

In the formulation of the equilibrium case, it was sufficient to specify state-occupation probabilities (p_i) and to give transition-equilibrium constants (j_i) for any arbitrarily chosen pair of states without considering whether a direct transition between these two states can occur or not. In the case of dynamic analysis, however, specifying individual transitions is unavoidable, and requires some additional considerations.

The first question concerns whether the assumption that the states in a conformational space are accessible to each other by all possible ways is reasonable or not. Such weblike connections in the conformational space are not only unnecessary, but also unlikely, because it is reasonable to expect the transitions to take place between alike states, but not to occur directly between very "distant" states. In other words, there must be a reasonable autocorrelation (i.e., "smoothness") in the process while traveling among the states. Otherwise, it would be difficult to explain different time-scale movements in protein molecules that have been observed experimentally (see for example Reference 51). Thus, if we assume that transitions are reversible (in the sense that if π_{ij} exists for any $i \neq j$, then π_{ji} also exists, with $\pi_{ij} \neq \pi_{ji}$, or $\pi_{ij} = \pi_{ji}$), then, in a conformational space of size n, one needs at least $n - 1$ connections (and $2(n - 1)$ rate constants) in order to confer strictly nonzero equilibrium probabilities for all the members (s_i) of the space (S). Let us call the set of such connections a "minimal path." Such minimal paths can be "unbranched" or "branched" depending on whether a state in the path is neighboring (or connected to) at most two states, respectively. Figure 4.4 shows an example of such minimal paths for $n = 4$.

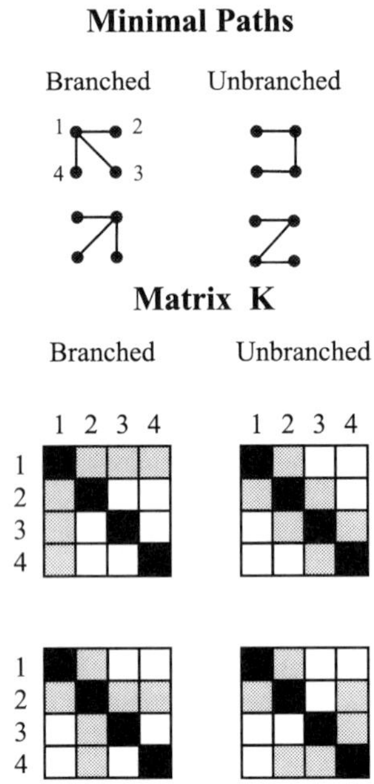

FIGURE 4.4 Examples of minimal paths (branched or unbranched) for a conformational space of size 4 (upper part of the figure). States are arbitrarily numbered from 1 to 4 and represented as small circles. Lines between states indicate possible transitions between states. **K** matrices (4×4 each) for the corresponding connectivities are schematically shown in the lower part of the figure: blacks are diagonal entries [Equation (10)], grays are off-diagonals ones, and whites are zero. Under these assumptions, **K**'s are always entry-wise symmetric. In the case of unbranched paths they contain at most two off-diagonal entries in the rows (or columns).

Any transition topology in the conformational space can be obtained by the superposition of such minimal paths. However, for the sake of simplicity, we shall consider only unbranched minimal paths here and provide some numerical results with $n = 100$.

Every permutation of state-indices defines an unbranched minimal path in the conformational space. Therefore, the total number of possible unbranched minimal paths is $n!$ for a conformational space of size n. However, with the assumption of reversibility at the connections, the number of distinguishable paths reduces to $n!/2$ (e.g., for $n = 4$ the path 1-3-2-4 is equivalent to its reverse (4-2-3-1), which are counted separately in $n!$). For $n = 100$, this number is on the order of 10^{157}.

With $n = 100$, such unbranched minimal paths consist of 99 connections, which require 198 (2×99) entries for **K**. In the following simulations, we assigned random numbers to the off-diagonal entries of **K** drawn from a Gaussian distribution (with a mean of 4 per unit time, corresponding to a time constant of $\approx$0.17 time unit). The range of such entries was 0.09 to 0.5 units in terms of time constants. Diagonal entries were then calculated according to Equation (10). The matrix **K** thus constructed has 100 different eigenvalues (one of which is zero) and the range of these eigenvalues is always much larger than that of the rate constants. A specific example

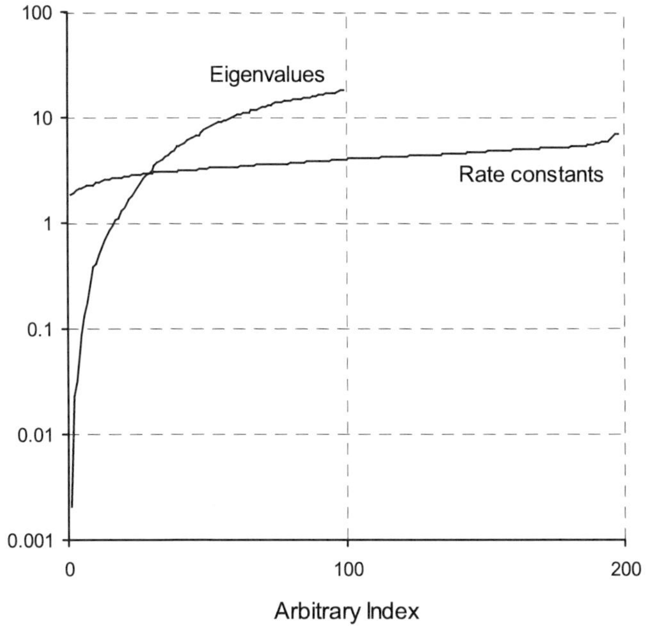

FIGURE 4.5 Comparative illustration of distributions of eigenvalues and rate constants for a random **K** matrix that corresponds to an arbitrary unbranched minimal path in a conformational space of size 100. Entries of **K** were randomly assigned from a Gaussian distribution with mean = 4 and a standard deviation of 1. Entries and eigenvalues of **K** are shown in the same (log) scale (ordinate). Both the entries and the eigenvalues are independently sorted in an assending order and indexed accordingly (abscissa).

is given in Figure 4.5 where the distribution of 99 nonzero eigenvalues and 198 rate constants (sorted in ascending order) are illustrated.

The results shown in Figure 4.5 imply that the process of traveling among states, $\mathbf{x}(t)$, is contributed by a great diversity of time constants compared to the individual state-transition rates. For instance, there is a 4-order-of-magnitude difference between the smallest and the largest eigenvalues, whereas the same difference for the transition rate constants is less than an order of magnitude. The significance of this result can be visualized by analyzing the full relaxation of the protein from an arbitrary initial configuration to a final one imposed by a given **K**. In Figure 4.6, we calculated the evolution of the state probabilities of a uniform distribution. The initial and final distributions shown in Figure 4.6 can be considered as the configuration of empty and ligated receptors, respectively.

It is evident (Figure 4.6) that the full relaxation takes about 1000 time units, while the average value of microscopic transition constants (4 per unit time) that we used in this simulation implies a time constant of 0.17 unit. Thus, the overall relaxation process of a protein exploring a large conformational space is very slow compared to the time scale of each transition rate between individual microstates. In order to show this in a more compact way we may define overall (ligand-induced) relaxation as the sum-of-squared differences between instantaneous and final (target)

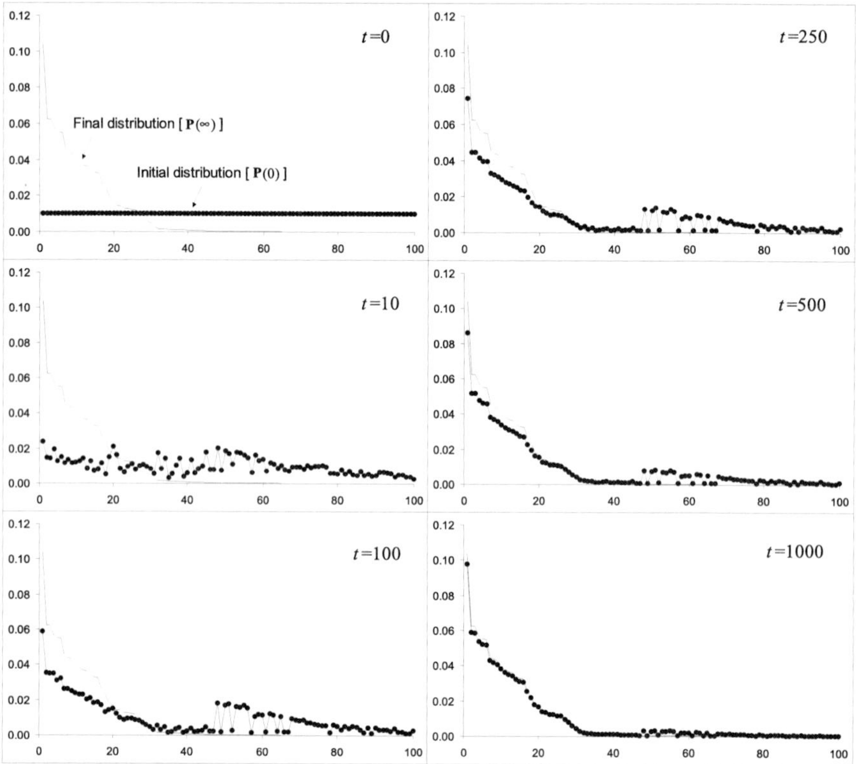

FIGURE 4.6 Snapshots of the evolution from an initial distribution to a final distribution in a conformational space of 100 elements. Ordinates are probabilities and abscissae are arbitrary state index. State index is sorted such that the state probabilities in the final distribution are arranged in a descending order. Initial distribution was assumed to be uniform (used to simulate empty receptor) as indicated in the upper-left panel (t = 0). Final distribution (used to simulate ligand-bound receptor) was imposed by the given matrix **K**. The same numerical values were used for the **K** matrix as were used in Figure 4.5. In order to visualize the convergence of the instantaneous distribution, the final distribution was showed in all panels (as indicated in the upper left panel). Equation (15) was used to calculate final distribution and Equation (14) to calculate instantaneous distributions at indicated time points.

distributions and plot it depending on time. In Figure 4.7 we showed such a relaxation using the simulation data presented in Figure 4.6.

In these specific examples (Figures 4.5, 4.6, and 4.7), the discrepency between time scales of individual microstate transitions and overall macroscopic relaxation is 3 orders of magnitude. However, it can be shown that these time scales can be potentially more divergent for a conformational space of the same size. In order to visualize this potential behaviour it may be helpful to further analyze the eigenvalues of **K**. As we mentioned above, the nonzero eigenvalues of **K** are the actual rate constants that contribute to the overall relaxation process. Thus, the smallest (in absolute terms) of these eigenvalues reflects the slowest rate that contributes to the

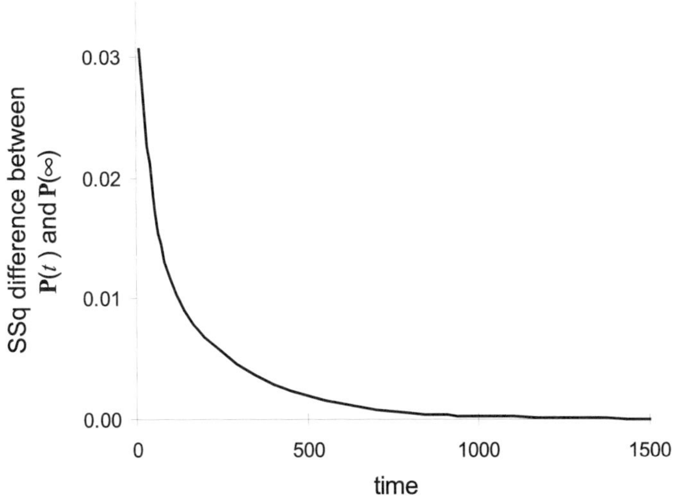

FIGURE 4.7 Sum of squared differences between instantaneous distribution, $\mathbf{P}(t)$, and final distribution, $\mathbf{P}(\infty)$ according to the multistate stochastic model for a conformational space of size 100. Initial distribution was arbitrarily assumed to be uniform (used to simulate empty protein). Final distribution (used to simulate ligand-bound receptor) was calculated using the same **K** as the one used in Figure 4.5 and 4.6. The curve can be seen as the overall conformational relaxation of receptor protein upon ligation by a conformationally "active" ligand. Full relaxation takes about 1000 time units, whereas the average rate constants used to construct **K** implies a time constant of 0.17 time units. Thus, there is a 3-order-of-magnitude difference between the average transition rates of individual microstates and macroscopic relaxation.

relaxation process. Figure 4.8 shows the distribution of the smallest nonzero eigenvalues calculated from 1000 different **K** matrices randomly constructed as above: i.e., each one of the analysed matrices corresponds to an arbitrary minimal-unbranched path and consists of random entries picked from a Gaussian distribution (mean = 4 per unit time). The distribution in Figure 4.8 is very skewed towards zero, implying that even slower rates compared to the specific example given in Figures 4.5 through 4.7, are possible (i.e., rates of up to 8 orders of magnitude slower as compared to that implied by the smallest rate constant).

All these examples were considered in a conformational space of size 100. As one might expect, however, the general discrepancy between the time scale of individual transitions and overall relaxation would increase with the number of states in the conformational space. Figure 4.9 shows the relationship between the size of the conformational space and macroscopic relaxation rates. It is evident from this simulation that: i) there is an inverse relationship between the number of states in the conformational space and the expected rates of the overall relaxation process, ii) if we assume individual transitions in the order of picoseconds (as in the molecular dynamic simulations), macroscopic relaxations with rates of minutes arise when there are ~10^4 states in the conformational space. With 10^6 states, the average contributions to the macroscopic relaxations is already on the order of minutes and even slower rates, (up to hours), become possible.

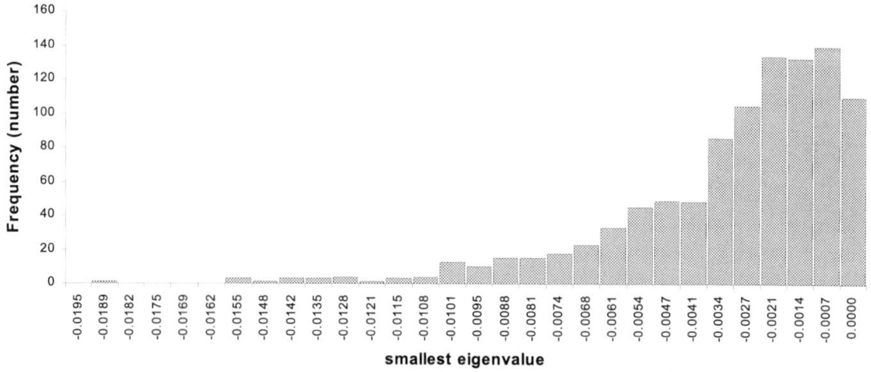

FIGURE 4.8 Distribution of smallest nonzero eigenvalues calculated from 1000 randomly constructed **K** matrices of rate constants for a conformational space of size 100. Entries of each **K** were picked from a normally distributed random variable with $\mu = 4$, $\sigma = 1$. **K**s were constructed in such a way that they correspond to minimal unbranched paths in the conformational space.

The picture in Figure 4.9 was obtained with unbranched-minimal paths, for which the macroscopic rates are expected to be slower compared to the nonminimal- or minimal-branched paths for a given numeric distribution of the entries of **K**. Thus, what we are guessing here is indeed the minimum number of states in the conformational space that confer such a discrepancy between individual transitions and the rates of overall relaxations. As the topology of the transition paths gets more complicated, one will obviously need more states to keep the discrepancy in these orders.

4.6 INITIAL-STATE-DEPENDENT MACROSCOPIC RELAXATION IN FLUORESCENTLY LABELED β_2-ADRENOCEPTORS

Here we shall consider an experimental example in the β_2-AR that can hardly be explained by the simple two-state allosteric model, but can be understood by the multistate-stochastic model discussed above.

The experimental data represent changes in fluorescence intensity upon ligand binding of fluorophore-labeled β_2-AR. The authors studied several ligands having different efficacy for adenylyl cyclase stimulation, and found that the change in fluorescence intensity was correlated with the efficacy of ligands. They concluded that the fluorophore(s) in the receptor molecule is sensitive to the conformational change that leads to receptor activation.[39,49] However the kinetics of the change in fluorescence intensity exhibited peculiar properties:

1. Isoproterenol (100 μM) and propranolol (1 μM) reduces and increases the fluorescence intensity, respectively, upon binding to the receptor. Full relaxation with either ligand takes about 15 minutes.

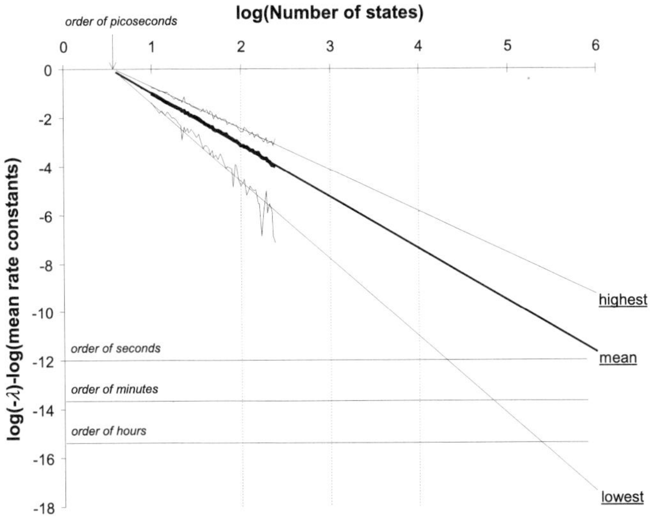

FIGURE 4.9 Relationship between (potential) rates of macroscopic relaxations and the size of conformational space. Possible contributions to macroscopic relaxation were shown as the distributional properties of eigenvalues of **K** at a corresponding number of states. Distributions of eigenvalues at conformational spaces of indicated size were calculated from randomly generated 500 **K** matrices that correspond to minimal-unbranched paths in the relevant conformational space (within the range of size n = 10 to 250). Lowest, highest, and mean values of eigenvalues thus calculated were shown against the number of states in a double-log plot. Numerical values of the entries of all **K** matrices used in the entire simulation were randomly assigned from the same normal distribution ($\mu = 4$, $\sigma = 1$). Eigenvalue scale was normalized to the mean value of the entires of **K** (which is arbitrarily assumed to be on the order of picoseconds as indicated in the picture). Note that the smaller the eigenvalues, the slower the processes. Straight lines indicate linear extrapolations of the simulation results towards higher (n > 250) or lower (n < 10) number of states. It is apparent that there is an inverse (log linear) relationship between the number of states in the conformational space and the contributing rates (as eigenvalues) to the overall relaxation process.

2. If, however, propranolol is applied in the presence of isoproterenol, relaxation in fluorescence intensity becomes an order of magnitude faster compared as to the one observed with propranolol alone.

Considering the concentrations of ligands used in the experiments, and the known binding rate constants of the ligands, it is not possible to explain the above phenomena by the binding kinetics of these ligands: i) binding time-constants predict much faster processes than those that have been observed in fluorescence experiments, and ii) the binding reaction of propranolol is expected to be slower (but not faster) in the presence of 100 μM isoproterenol. Thus, it appears that the kinetics of the fluorescence change are not limited by the rate of ligand binding to the receptor. Therefore, it can be assumed that the observed change in fluorescence intensity reflects the rate of ligand-induced conformational change in the protein.

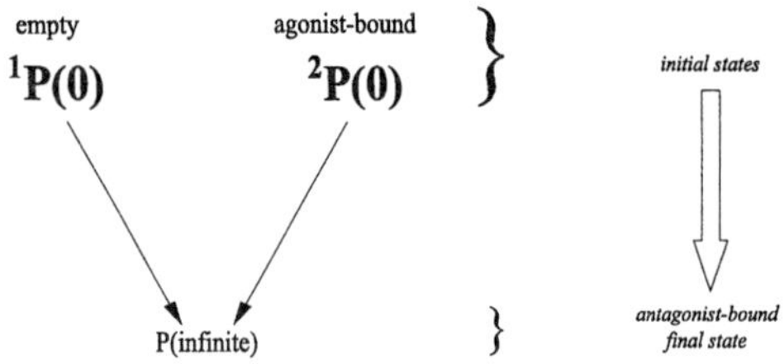

FIGURE 4.10 Schematic representation of the experimental design where the relaxation of two different initial states of the receptor protein (empty or agonist-stimulated) to a common final state (antagonist-bound) were observed. These macroscopic states, in the language of the model, are indicated as $^1\mathbf{P}(0)$, $^2\mathbf{P}(0)$, and $\mathbf{P}(\mathit{infinite})$, respectively. Each **P** corresponds to a different equilibrium probability distribution over the same conformational space (see text for details).

Why does the rate of conformational change induced by the same ligand (propranolol) differ depending on whether we start with an empty or with an agonist-bound receptor? Let us first translate the question into the language of the stochastic model. Suppose we have two ligands (e.g., an agonist and an antagonist), and we wish to compare the following two processes: i) The movement of the protein from the empty state to the antagonist bound state, and ii) from the agonist bound state to the same antagonist bound state (as schematized in Figure 4.10).

Both processes have a common equilibrium target (i.e., a constant **K**), since they point to the same antagonist-bound macroscopic state of the receptor, hence, they imply two relaxation processes with the same set of eigenvalues but different initial distributions, i.e., $^1\mathbf{P}(0)$ or $^2\mathbf{P}(0)$, which may be assumed to correspond to empty or agonist-bound forms of the receptor. We may further define the overall relaxation again as the sum-of-squared differences between the instantaneous occupation probabilities, $p_i(t)$, of microstates, s_i, and the final equilibrium probabilities of the corresponding microstates, $p_i(\infty)$, as we did in the previous section.

Let us first consider the two-state case, which is exactly analogous to the two-state allosteric model. The conformational space in this case consists of two elements, s_1 and s_2, and the corresponding transfer matrix **K** reads as

$$\mathbf{K} = \begin{bmatrix} -k_{12} & k_{12} \\ k_{21} & -k_{21} \end{bmatrix}$$

K has two eigenvalues: $\lambda_1 = 0$, $\lambda_2 = -(k_{12} + k_{21})$, and according to Equation (14) the instantaneous state probabilities, $p_1(t)$ and $p_2(t)$, are given as

$$\begin{aligned} p_1(t) &= p_1(\infty) + [p_1(0) - p_1(\infty)]e^{\lambda_2 t} \\ p_2(t) &= 1 - p_1(t) \end{aligned} \tag{16}$$

Although the degree of freedom is 1 in this special case, (i.e., $p_2(t)$ is fully determined by $p_1(t)$, or vice versa) we still can use the above definition of macroscopic relaxation as the sum-of-squared differences of instantaneous and target probabilities, for the sake of consistancy:

$$\Delta(t) = [p_1(t) - p_1(\infty)]^2 + [p_2(t) - p_2(\infty)]^2$$

Substitutions from Equation (16) and a rearrangement yields:

$$\Delta(t) = 2[p_1(0) - p_1(\infty)]^2 e^{2\lambda_2 t}$$

Normalizing the latter equation with respect to the initial value of the sum of squared deviations (which is a constant for a given case) yields:

$$\Delta_n(t) = e^{2\lambda_2 t} \tag{17}$$

which is independent of the initial distribution of the two states. Thus, as intuition would predict, the rigorous two-state model does not allow the rate of the overall relaxation to depend on the initial configuration of the protein. However, as we mentioned above, the experiments in β_2-AR have shown otherwise: The rate of fluorescence relaxation induced by propranolol is different depending on whether the receptor is empty or isoproterenol-bound.

Next, we show that such a phenomenon can be explained by the multistate stochastic model. Again, we shall use numerical simulations.

In these simulations we used the same matrix **K** as in Figures 4.5 through 4.7. **K** corresponds to an arbitrary minimal-unbranched path in a conformational space of size 100, and its entries were assigned randomly from a normal distribution of $\mu = 4$, $\sigma = 1$. The final distribution of microstates in the conformational space is determined by **K** alone [see Equation (15)], since the elements of columns of $\mathbf{A}_1$ are always equal and $\Sigma pi(0) = 1$ by definition, which makes the resulting final probabilities, $\mathbf{P}(\infty)$, independent of the initial distribution $\mathbf{P}(0)$. Therefore, a constant **K** always implies a constant final equilibrium distribution regardless of $\mathbf{P}(0)$. Let us then assume that the final distribution imposed by the constant **K** corresponds to the distribution of the microstates when the antagonist is bound to the receptor. Thus, by analyzing the relaxation of the state distributions from different initial configurations to this final distribution (see Figure 4.10), we can simulate the experimental results obtained in the fluorescence experiments of the β_2-AR. Figure 4.11 shows the results of such a simulation as snapshots of the state distributions from different time points, where three different initial distributions (i.e., a uniform, a triangular, or a Gaussian distribution) relax to a common final distribution imposed by a constant **K**. Note that overall relaxations appear to be different depending on the initial distribution. In order to visualize this difference, we further calculated overall relaxations of the three cases as the normalized sum-of-squared differences between instantaneous and final probabilities, as we did in Equation (17). Results are shown in Figure 4.12, where considerable differences in relaxation times are evident depending on the initial distribution. These results can be translated back to the

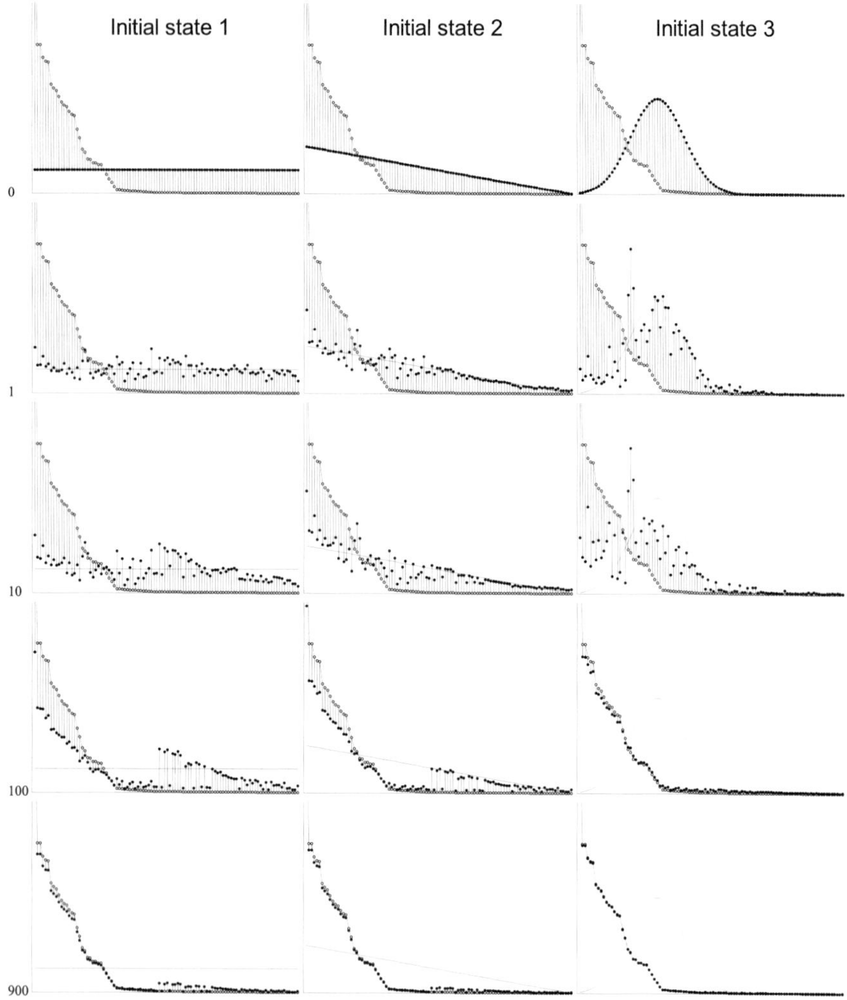

FIGURE 4.11 Snapshots from relaxations of three different initial distributions to a common final distribution in a conformational space of size 100. Abscissa and ordinate of each panel indicate arbitrary state index and probability, respectively. Initial distributions are shown in the first row (t = 0) and indicated as initial state 1, 2, and 3. Each column shows the evolution of corresponding initial distribution. Numbers at the left-hand side indicate the time points at which the snapshots are taken. Filled and open symbols indicate instantaneous and final probabilities, (i.e., $\mathbf{P}(t)$ and $\mathbf{P}(\infty)$), respectively. The area between instantaneous and final distributions is shaded to facilitate the perception of overall relaxations in each case. Dotted lines in each panel show traces of initial distributions. The same **K** matrix was used for all the three cases, which confers a common final distribution. Numerical values of the entries and the connectivities of the **K** matrix are as in Figure 4.2 through 4.4 (i.e., it corresponds to an arbitrary unbranched minimal path in the conformational space of size 100, and its entries are randomly assigned from a Gaussian distribution of $\mu = 4$ (per unit time) and $\sigma = 1$). Instantaneous probabilities for given **K** and initial distribution were calculated by using Equation (14).

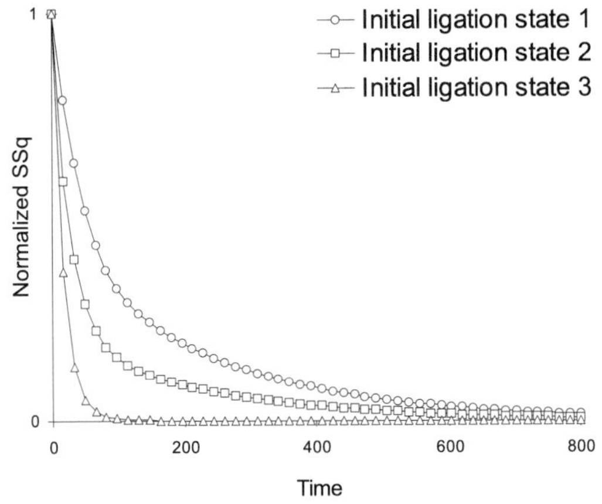

FIGURE 4.12 Overall relaxation of three different initial states to a common final state. The same data in Figure 4.8 are represented as the normalized sum-of-squared differences between instantaneous and final probabilities (i.e., $\Sigma[p_i(t) - p_i(\infty)]^2$ divided by $\Sigma[p_i(0) - p_i(\infty)]^2$). Indicated initial ligation states 1, 2, and 3 correspond to the uniform, triangular, and Gaussian distributions, respectively, shown in Figure 4.8. Note that the curves actually reflect the evolution of the shaded areas in Figure 4.8.

experimental language as follows: We are observing a conformational relaxation induced by an antagonist ligand (measured as relaxation of fluorescence intensity), which is slow when the ligand is applied to the empty receptor, and becomes faster when it is applied to the agonist-bound receptor. This is also the case in the simulations when we arbitrarily assume that an empty receptor is identified by a uniform distribution, and agonist-bound receptor by the Gaussian distribution (indicated as initial states 1 and 3, respectively, in Figures 4.11 and 4.12).

4.7 CONCLUSION

We classified models of receptor action at the molecular level in two main categories:

1. Ternary-complex-like models, where the conformational state of the receptor protein is not considered explicitly.
2. Allosteric models, where one-to-one correspondence between conformational and functional states of the protein is assumed.

The probabilistic model of receptor function discussed above represents a new category, where functional states are the average result of the intramolecular motion of the protein among an indefinite number of conformational states. The relative value of such models largely depends on the types of experimental questions that we are posing. As we try to map macroscopic function onto detailed physical properties of the receptor (or we engineer single residue mutations to modify function), we move from the first category toward the last.

We recently discussed experimental observations of the steady-state behaviour of GPCR that cannot be explained by the models in the first two categories, but can be understood with the multistate-probabilistic model.[39] Here, we exemplified kinetic properties, which again requires the consideration of a large conformational space at nonstationary conditions, and shows that the probabilistic model at nonequilibrium states can be derived using Markoff assumptions.

The probabilistic model discussed above can be seen as the description of the protein at a very microscopic level. On the other hand, the model is scalable to the ultimate macroscopic level of the ternary complex model, because there are precise links between the parameters of these two models [see for instance Equations (1) and (2)]. Therefore, the probabilistic model can be considered as a simple tool to bridge structural and dynamic properties of the protein (stationary or non-stationary) to its macroscopic molecular function.

REFERENCES

1. Frauenfelder, H., Parak, F., and Young, R. D., Conformational substates in proteins, *Ann. Rev. Biophys. Biophys. Chem.,* 17, 451, 1988.
2. Frauenfelder, H., Sligar, S. G., and Wolynes, P. G., The energy landscapes and motions of proteins, *Science,* 254, 1598, 1991.
3. Wyman, J., The binding potential, a neglected linkage concept, *J. Mol. Biol.*, 11, 631, 1965.
4. Wyman, J., Allosteric linkage, *J. American Chem. Soc.*, 89, 2202, 1967.
5. Del Castillo, J. and Katz, B., Interaction at end-plate receptors between different choline derivatives, *Proc. R. Soc. Lond. (Biol. Sci.),* 146, 369, 1957.
6. Monod, J., Wyman, J., and Changeux, J. P., On the nature of allosteric transitions: A plausible model, *J. Mol. Biol.*, 12, 88, 1965.
7. Zwanzig, R., Two-state models of protein folding kinetics, *Proc. Natl. Acad. Sci.,* 94, 148, 1997.
8. Rodbell, M., The role of hormone receptors and GTP-regulatory proteins in membrane transduction, *Nature,* 284, 17, 1980.
9. Gilman, A. G., G Proteins: transducers of G protein generated signals, *Ann. Rev. Biochem.*, 56, 615, 1987.
10. Neer, E. J. and Clapham, D. E., Role of G protein subunits in transmembrane signalling, *Nature,* 333, 129, 1988.
11. Ross, E. M., Signal sorting and amplification through G protein-coupled receptors, *Neuron,* 3, 141, 1989.
12. Brinbaumer, L., G proteins in signal transduction, *Ann. Rev. Pharmacol. Toxicol.*, 30, 675, 1990.
13. Sprang, S. R., G protein mechanisms: insights from structural analysis, *Ann. Rev. Biochem.*, 66, 639, 1997.
14. Bourne, H. R., How receptors talk to trimeric G proteins, *Curr. Opin. Cell. Biol.*, 9, 134, 1997.
15. Clark, A. J., *The Mode of Action of Drugs on Cells*, Edward Arnold and Co., London, 1933.
16. Ariëns, E. J., Affinity and intrinsic activity in the theory of competitive inhibition, *Arch. Int. Pharmacodyn. Ther.,* 99, 32, 1954.

17. Ariëns, E. J., Simonis, A. M., and Van Rossum, J. M., Drug receptor interaction: interaction of one or more drugs with one receptor system, in *Molecular Pharmacology: The Mode of Action of Biologically Active Componds*, E. J. Ariëns, Ed., Academic Press, New York, 1964, 119.
18. Gaddum, J. H., Theory of drug antagonism, *Pharmacol. Rev.*, 9, 21, 1957.
19. Stephenson, R. P., A modification of receptor theory, *Br. J. Pharmacol.*, 11, 379, 1956.
20. Van Rossum, J. M., The relationship between chemical structure and biological activity, *J. Pharm. Pharmacol.*, 15, 285, 1963.
21. Karlin, A., On the application of "a plausible model" of allosteric proteins to the receptor for acetylcholine, *J. Theor. Biol.*, 16, 306, 1967.
22. Rang, H. P., Receptor Mechanisms, *Br. J. Pharmacol.*, 48, 475, 1973.
23. Thron, C. D., On the analysis of pharmacological experiments in terms of an allosteric receptor model, *Mol. Pharmacol.*, 9, 1, 1973.
24. Colquhoun, D., The relationship between classical and cooperative drug action, in *Drug Receptors*, H. P. Rang, Ed., University Park Press, Baltimore, 1973, 149.
25. Black, J. W. and Leff, P., Operational models of pharmacological agonism, *Proc. R. Soc. Lond. (Biol. Sci.)*, 220, 141, 1983.
26. Wyman, J. and Gill, S. J., *Binding and Linkage: Functional Chemistry of Biological Macromolecules*, University Science Books, Mill Valley, California, 1990.
27. Weber, G., *Protein Interactions*, Chapman and Hall, New York, London, 1992.
28. Costa, T., Ogino, Y., Munson, P. J., Onaran, H. O., and Rodbard, D., Drug efficacy at guanine nucleotide-binding regulatory protein-linked receptors: thermodynamic interpretation of negative antagonism and of receptor activity in the absence of ligand, *Mol. Pharmacol.*, 41, 549, 1992.
29. Onaran, H. O., Costa, T., and Rodbard, D., βγ-Subunits of guanine nucleotide-binding proteins and regulation of spontaneous receptor activity: thermodynamic model for the interaction between receptors and guanine nucleotide-binding protein subunits, *Mol. Pharmacol.*, 43, 245, 1993.
30. Onaran, H. O. and Costa, T., Agonist efficacy and allosteric models of receptor action, *Ann. NY Acad. Sci.*, 812, 98, 1997.
31. Leff, P., The two-state model of receptor activation, *Trends Pharmacol. Sci.*, 16, 89, 1995.
32. Samama, P., Cotecchia, S., Costa, T., and Lefkowitz, R. J., A mutation-induced activated state of the β2-adrenergic receptor, *J. Biol. Chem.*, 268, 4625, 1993.
33. De Lean, A., Stadel, J. M., and Lefkowitz, R. J., A ternary complex model explains the agonist-specific binding properties of the adenylate cyclase-coupled β-adrenergic receptor, *J. Biol. Chem.*, 255, 7108, 1980.
34. Ehlert, F. J., The relationship between muscarinic receptor occupancy and adenylate cyclase inhibition in the rabbit myocardium, *Mol. Pharmacol.*, 28, 410, 1985.
35. Weber, G., Ligand binding and internal equilibria in proteins, *Biochemistry*, 11, 864, 1972.
36. Wregget, K. A. and DeLean, A., The ternary complex model: Its properties and application to ligand interaction with the D2-dopamine receptor of the anterior pituitary gland, *Mol. Pharmacol.*, 26, 214, 1984.
37. Weiss, J. M., Morgan, P. H., Lutz, M. W., and Kenakin, T. P., The cubic ternary complex receptor-occupancy model I. model description, *J. Theor. Biol.*, 178, 151, 1996.
38. Gill, S. J., Richey, B., Bishop, G., and Wyman, J., Generalized binding phenomena in an allosteric macromolecule, *Biophys. Chem.*, 21, 1, 1985.

39. Onaran, H. O., Scheer, A., Cotecchia, S., and Costa, T., A look into receptor efficacy. From the signalling network of the cell to the intramolecular motion of the receptor, in *General Pharmacology: The Pharmacology of Functional, Biochemical, and Recombinant Receptor Systems*, Handbook of Experimental Pharmacology, T., Kenakin and J. A., Angus, Eds., Springer Verlag, Heidelberg (in press).
40. Tolman, R. C., *The Principles of Statistical Mechanics*, Oxford University Press, New York, 1938.
41. Papoulis, A., *Probability, Random Variables, and Stochastic Processes*, McGraw-Hill, Singapore, 1984.
42. Colquhoun, D. and Hawkes, A. G., Relaxation and fluctuation of membrane currents that flow through drug-operated channels, *Proc. R. Soc. Lond. B.*, 199, 231, 1977.
43. Colquhoun, D. and Hawkes, A. G., The principles of stochastic interpretation of ion-channel mechanisms, in *Single Channel Recording*, B. Sakmann and E. Neher, Eds., Plenum Press, New York, 1983, 135.
44. Brooks, C. L., Karplus, M., and Pettit, B. M., Proteins: A theoretical perspective of dynamics, structure and thermodynamics, in *Advances in Chemical Physics*, Vol. 71, I. Prigogine and S. A. Rice, Series Eds., John Wiley & Sons, New York, 1987.
45. Bryngelson, J. D., Onuchic, J. N., Socci, N. D., and Wolynes, P. G., Funnels, pathways, and energy landscape of protein folding: a synthesis, *PROTEINS:Structure, Function and Genetics,* 21, 167, 1995.
46. Wittung-Stafshede, P., Lee, J. C., Winkler J. R., and Gray H. B., Cytochrome b562 folding by electron transfer: Approaching the speed limit for formation of four-helix-bundle protein, *Proc. Natl. Acad. Sci. USA,* 96, 6587, 1999.
47. Karplus, M. and Petsko, G. A., Molecular dynamics simulations in biology, *Nature,* 347, 631, 1990.
48. Agmon, N. and Hopfield J. J., CO binding to heme proteins: A model for barrier height distributions and slow conformational changes, *J. Chem. Phys.*, 79, 2042, 1983.
49. Gether, U., Lin, S., and Kobilka, B. K., Fluorescent labeling of purified β2 adrenergic receptor. Evidence for ligand-specific conformational changes, *J. Biol. Chem.*, 270, 28268, 1995.
50. Auzat, I., Gawlita E., and Garel, J. R., Slow ligand-induced transitions in the allosteric phosphofructokinase from Escherichia coli, *J. Mol. Biol.*, 249, 1995.
51. Gerstein, M., Lesk, A. M., and Chothia, C., Structural mechanisms for domain movements in proteins, *Biochemistry,* 33, 6739, 1994.

5 Linking Bioinformatics and Cheminformatics: From Genes to Structure-Activity Relationships

Michael W. Lutz

CONTENTS

0-8493-0100-9/01/$0.00+$.50

5.1 INTRODUCTION

The terms bioinformatics and cheminformatics refer to the use of computational methods in the study of biology and chemistry. Information derived from DNA or protein sequences, protein structure, and chemical structure is used to build models of biochemical systems or models of the interaction of a biochemical system with a small molecule. There are mathematical and statistical methods for analysis, public databases, and literature associated with each of these disciplines. However, there is substantial value in considering the interaction between these areas and in building computational models that integrate data from both sources. In the most general sense, integrating bioinformatics and cheminformatics leads to models that relate features of biological systems (sequences, protein structures, motifs) to features of the chemical structures, including small organic molecules, that interact with them.

This chapter discusses computer modeling that links bioinformatics and cheminformatics. The first part of the chapter describes the types of information that are available for use in the computational models, both derived from sequence and protein databases and from calculations based on chemical structure. The use of gene expression data in quantitative models is also discussed. The remainder of the chapter describes several different types of computational models: visualization and clustering, hidden Markov models, neural networks, and classification and regression trees. Several case studies that specifically address questions of how one can relate aspects of the interaction of biochemical targets (proteins or receptors) with features of chemical structures are used to show how the models utilize information from bioinformatics and cheminformatics. Other case studies are concerned with questions of classification and grouping: this information is useful for drug design in order to identify pharmacophores. Some of the case studies make direct use of data calculated from proteins and small organic molecules. An appendix to the chapter contains a listing of bioinformatics databases and several references for software.

5.2 BIOINFORMATICS DATABASES: CONTENT AND ISSUES RELATED TO INTEGRATION OF DATA

The World Wide Web greatly facilitates mathematical modeling using data from public databases. Large public databases of nucleotide sequences, protein sequences and structures, and literature references about specific genetic sequences or protein structures are all available on the web. Results from the human genome project are posted on the web along with full annotations. Software tools for visualization and analysis of genetic and genomic data are also available on the web from universities and centers including the National Center for Biotechnology Information (NCBI) and the European Molecular Biology Institute (EMBI).

Bioinformatics data is found in literally hundreds of different databases and new databases become available frequently. Investigators working with a specific model organism (e.g., c.elegans, drosophila, zebrafish, yeast), human data or system (ion channel, or 7-transmembrane (7TM) receptor) should search the web and recent literature for new databases in addition to the major databases listed here. When using any database, it is important to recognize that there will likely be errors in the data (misclassification of genes or proteins, missing data, missing links to other data) and that at least at present, integrating data from different sources is largely a manual process. The tools associated with the databases, e.g., Entrez and the Sequence Retrieval System (SRS), do some integration, primarily by making links between literature references and gene or protein identifiers. Issues related to integration of data and errors in the databases are discussed in a paper by Macauley et al.[1]

There are several major databases that are a good starting point for investigators: for nucleic acid sequences, GenBank (the International Nucleic Acid Sequence Data Library); for protein sequences, SWISS-PROT; for protein structures, Brookhaven's protein databank or PDB; and for the structures of nucleic acid sequences, the Nucleic Acid Database or NAD. Many of these databases have web-based software available to extract and analyze data. For example, Entrez and BLAST are available from the NCBI web page to perform text searches and sequence similarity searches of the GenBank database and other nucleotide and protein databases including SWISS-PROT. The Sequence Retrieval System (SRS) is available for searching databases supported by the European Molecular Biology Laboratory (EMBL). In addition to these databases, there are numerous databases organized by organism. The WWW Virtual Library is a good starting place to find data categorized by organism.

The databases listed above are excellent resources for finding information about specific sequences and structures, for identifying contacts who have worked with the sequences and structures, and for searching for related sequences and structures. These are good databases for models that require sequence or structure as input. For computational modeling, derived databases are often more useful for providing motifs and patterns based on structure. Derived databases are constructed by analysis, restructuring, or additional annotation of existing data. Examples include the BLOCKS database that is based on conserved regions of proteins, the protein motif database (PROSITE), and the structural classification of proteins (SCOP).

Information about gene expression is available from databases designed to handle this type of information. The gene expression databases are organized by species (including human, c. elegans, zebrafish, and mouse), organ, or class of protein. Although there are differences in structure between the databases, most contain information about the identification of the gene, expression patterns (both temporal and spatial), assay methods, and literature references. Many are comprehensive databases including information about genetic studies, physical maps, and other genomic information. Gene expression data is useful for constructing biological models that are investigated by computer simulation and experimentation.

For mathematical and statistical models, descriptors based on amino acid sequence or protein structure are calculated in a manner similar to that used with small organic molecules (vide infra). In addition to choice of descriptors, investigators must decide the length and locations of fragments of the sequence used for the calculations or use the full-length sequence. The descriptors are based on the characteristics of the amino acids including isotropic surface area and charge.[2-4] For many applications, principal component analysis is used to reduce the descriptor set from several hundred variables to a small number of orthogonal scores.

5.3 PROPERTIES AND DESCRIPTORS OF SMALL ORGANIC COMPOUNDS USED IN MODELS

Descriptors of small organic compounds are computed from the 2D representation of the chemical structure or from one or more 3D conformations. Properties of molecules (solubility, lipophilicity) are either measured in laboratory experiments or are estimated using computational models. Sets of descriptors and properties are the starting point for mathematical models that relate chemical structure to biological activity or attempt to cluster compounds based on their biochemical profile.

1D descriptors are scalars estimated from the 2D chemical structure: examples include molecular weight and calculated log P. Topological descriptors characterize the connections in the chemical graph and are often expressed as vectors.[5] These descriptors include atom counts, path lengths between atoms, and other measures obtained from the 2D graph. Descriptors often contain information about chemical reactivity (e.g., number of H-bond donors and acceptors). Fragment-based descriptors note the presence or absence of substructures in the molecule. Chemical database systems including MDL, Daylight, and TRIPOS contain computer code to calculate many different fragment-based descriptors. For some applications, the presence or absence of fragments is coded in a binary bitstring for each molecule.

Descriptors based on the 3D structure of a molecule require choice of a conformation or a process to combine results across different conformations. Often, the lowest energy conformations are used. However, there is no assurance that the low energy conformations chosen correspond to the protein-bound conformation or that a set of conformers is representative of the conformational space of a compound. Three-dimensional descriptors can be calculated as continuous variables, counts, or used to indicate presence and absence of structural features. If distance in the 3D space is included, descriptors based on combinations of features present in the 3D structure can be calculated. Table 5.1 shows a set of chemical feature types that are used to define 3D pharmacophores for a process of automatic pharmacophore identification as described by Chen et al.[6]

Analytical chemistry methods are used to measure chemical properties including hydrophobicity and solubility. These methods often require HPLC or mass spectroscopy for measurement. A catalogue of chemical properties and methods to measure them is given in Lyman et al.[7]

5.4 GENE EXPRESSION DATA AND ITS USE IN MATHEMATICAL MODELS

The databases described in the previous sections provide the data for numerous computational models for applications including prediction of gene structure, sequence analysis, and classification according to structure and/or function. Theoretically, the descriptors of proteins (e.g., isotropic surface area, amino acid

TABLE 5.1
Chemical Feature Types Used to Define Three-Dimensional Pharmacophores for a Process of Automatic Pharmacophore Identification. Reprinted with permission from Reference 6.

Types	Description
Negative charge center	carboxylic group, sulfinic group, phosphinic group, etc.
Positive charge center	nitrogen in primary, secondary, and tertiary amines, etc.
Hydrogen bond acceptor	nitrogen, oxygen, and sulfur with at least 1 available lone pair electron
Hydrogen bond donor	nitrogen, oxygen, sulfur, or the terminal of a triple bond linked with hydrogen
Aromatic ring center	center of aromatic 5- or 6-membered ring
Hydrophobic center	center of fragment containing more than 3 hydrophobic atoms
Nitrogen	nitrogen atom
Oxygen	oxygen atom
Sulfur	sulfur atom
Phosphor	phosphorus atom
Fluorine	fluorine atom
Halogen	chlorine atom, bromine atom, and iodine atom

content, charge) can be used in any mathematical model and some examples are described in later parts of this chapter. Quantitative gene expression data augments these data sources and is valuable in the development of computational models of cellular systems.

In order to be complete, models of cellular processes based on gene expression must include interactions: between the proteins encoded by genes, between proteins and endogenous hormones, and between the proteins and the environment. Computational methods are needed both to translate raw data into a quantitative readout (e.g., image analysis of differences in gene expression) and to construct models that relate gene or protein expression to function. Genetic regulation is a complex control process; groups of genes are involved both directly and indirectly in many different signaling processes. The tools of genomics are used to map genes to their products (usually proteins) and to sequence, and to analyze data about the distribution of gene products. Generally, messenger RNA (mRNA) is used to analyze gene expression. By analyzing mRNA from different organs, tissues, and cell types, the distribution of the products of a set of genes can be quantified.

Experimental methods to obtain gene expression data include hybridization assays using arrays of cDNA clones[8-11] or arrays of oligonucleotides,[12] serial analysis of gene expression,[13] and differential display.[14]

The principles behind glass chip-based gene expression assays[9] are shown in Figure 5.1. Total mRNA is isolated from a tissue or other sample. Reverse transcription in the presence of fluorescent nucleotides is done to label the mRNA for detection. The resulting mixture, also known as a probe, is then hybridized to cDNA that is linked to a glass slide, washed at a high stringency, and scanned for the fluorescence signal that will be proportional to the original expression level. Array

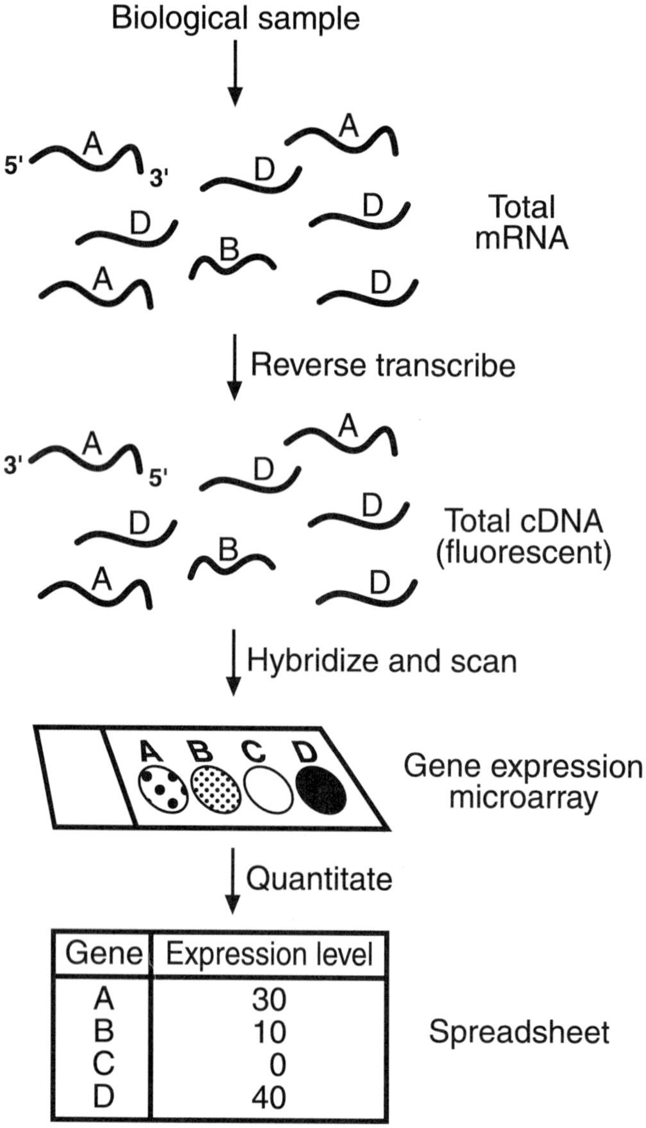

FIGURE 5.1 Chip-based gene expression assays. This diagram shows the steps to quantitate gene expression on a cDNA microarray. Total messenger RNA (mRNA) is isolated from a biological sample, primed with oligo-dT, and fluorescently labeled using a single round of reverse transcription in the presence of fluorescein-12-dCTP. The fluorescent probe is hybridized to a cDNA microarray containing specific hybridization targets, washed at high stringency, and scanned for fluorescence emission following laser excitation. Measurement of fluorescence intensity is used to quantitate gene expression. Reprinted with permission from Reference 9.

densities of greater than 1000 cDNAs per square centimeter make it possible to test against large portions of a genome; a 10 cm^2 array would contain cDNA for 10,000 different genes. The source of the cDNA can be fully sequenced clones, randomly chosen cDNA, or expressed sequence tags. The probe is either a complex mixture of cDNA fragments generated from the mRNA or genomic DNA. Biological samples can be compared in the same assay using two-color fluorescence detection.[9] Using two-color detection, expression in different tissues can be compared without separate hybridization steps and image subtraction. The cDNA on the chip can be arranged to support kinetic studies to show the expression data over multiple time points.

The volume of data from kinetic gene expression experiments is large; millions of observations are taken from tens of thousands of genes. Image analysis software is used to handle the image processing operations, quantify results, and compare images. Interpretation of the data can be as simple as point by point comparisons of expression in normal and diseased tissues, or as complex as mathematical models for clustering and detection of interactions.

Gene expression analysis is an important step in constructing models of genetic control and regulatory networks. Examples include the lysis-lsogeny decision in the bacteriophage lambda[15] and the diauxic shift from anaerobic growth to aerobic respiration in yeast.[16] The diauxic shift is a consequence of the depletion of glucose. In the yeast study, monitoring was at the whole genome level. DeRisi[16] developed an effective visual display for correlating changes in gene expression with changes in the metabolic cycle (Figure 5.2). The yeast genes associated with each step are identified on a biochemical pathway diagram. By color-coding the diagram, it is easy to see which genes have higher expression levels after the diauxic shift. There was significantly more expression for genes coding for aldehyde dehydrogenase, pyruvate carboxylase, and acetyl-coenzyme A synthase. There is less expression of transcription of genes encoding pyruvate decarboxylase. Biochemically, these changes rechannel pyruvate away from acetaldehyde to oxalacetate where it supplies the TCA cycle and gluconeogenesis.

For computational models of the cell, the kinetics of gene expression are of importance. Figure 5.3 shows the change in expression levels of several classes of genes associated with specific proteins in the TCA cycle pathway over time after depletion of glucose (e.g., after the switch to aerobic respiration). The average expression level of several genes is used in the graph to model the cumulative response of all the genes in a specific group and correlate expression against the decrease in glucose. There is a decrease in expression of genes associated with protein synthesis (ribosomal proteins, tRNA synthetases) and an increase in expression of genes associated with carbohydrate storage (cytochrome-c, TCA/glyoxalate cycle) as the levels of glucose decrease below 15 g/liter.

Using DNA microarrays to monitor differences in transcription provides an approach to analyze regulatory pathways and networks. Changes in genes serve as markers for the activity of drugs and can be used to screen for drugs that are associated with a specific expression pattern. Gene expression data, in combination with other experimental data about protein expression, enzyme kinetics, and cellular biochemistry provides the information necessary to construct models of cellular

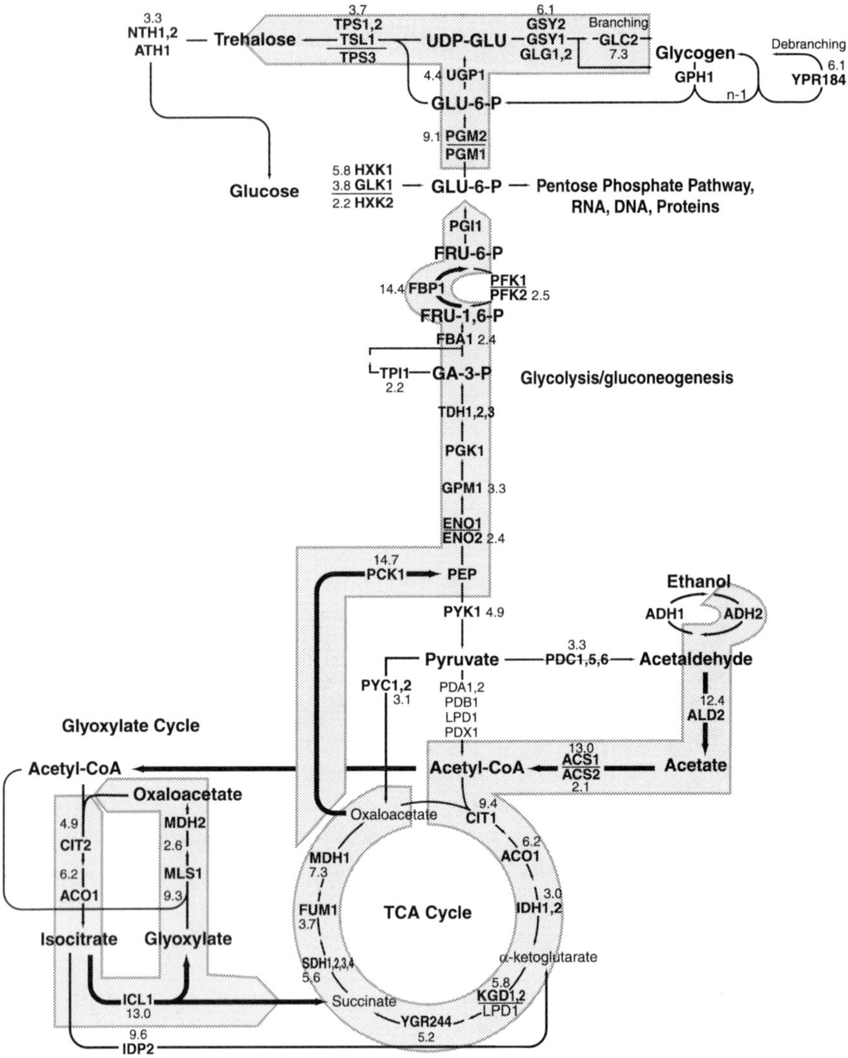

FIGURE 5.2 Gene expression data overlaid on metabolic pathway data. Correlation of gene expression data with the metabolic cycle in yeast. This figure shows part of the yeast metabolic pathway associated with carbon and energy metabolism. Changes in gene expression for the mRNA encoding the enzymes are overlaid on the pathway. Yeast genes encoding the enzymes are identified by the different codes in the boxes. Color or shading is used to indicate genes where expression levels change in the diauxic shift, either increased or decreased expression. The magnitude of the change is shown using the small numerals adjacent to the boxes. For multimeric enzyme complexes, the number is a unweighted average of all of the genes in the box. Reprinted with permission from Reference 16.

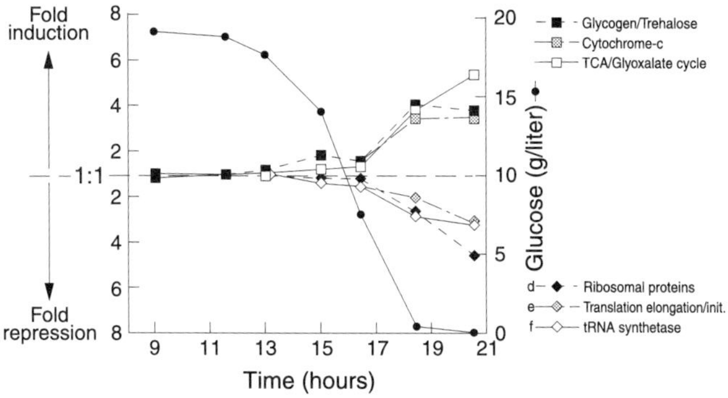

FIGURE 5.3 Quantitative expression of gene expression data. This figure shows the kinetics of gene expression for groups of functionally related genes. The curves represent the average induction or repression ratios for all of the genes in each indicated group. The number of genes in each group ranged from 15 (glycogen and trehalose synthesis and degradation) to 112 (ribosomal proteins). This figure also shows glucose concentration kinetics overlaid with the gene expression data. Abscissa: Time in hours. Ordinate: Average fold induction or repression ratio (left axis) or glucose concentration (g/liter) (right axis). Reprinted with permission from Reference 16.

regulatory networks. In many studies, the model is expressed as a schematic of the regulatory network where the kinetics of branches of the network form the next level of detail.

5.5 CASE STUDY 1: COMPUTER SIMULATION OF CELLULAR REGULATORY NETWORKS

Models of regulatory networks are translated into systems of differential equations for computer simulation. Cellular signaling involves complex relationships between different components: the enzymes and substrates of the chemical reactions. These reactions change over time and as a consequence of spatial compartmentalization in the cell. The importance of time lags, concentrations of components, and interactions is understood at a quantitative level by changing the model parameters and conducting experiments to test the results of the simulations. Experimental data is used to test and refine the mathematical models.

Novak and Tyson[17] illustrated the use of a differential-equation-based model for a cellular signaling pathway. Their model is of M-phase control in Xenopus oocyte extracts and intact embryos. The starting point for their model is a schematic of the M-phase control system (Figure 5.4); typical of the type of model of a cellular network that was described in the previous section based on gene and protein expression patterns and kinetics. Major variables in the model include the phosphorylation state of M-phase promoting factor (MPF), a protein kinase that is a heterodimer of cyclin, and Cdc2. The phosphorylation states of these subunits are in turn regulated by two kinase-phosphatase pairs: at the inhibitory phosphorylation site, a kinase (Wee-1) and a phosphatase (Cdc25), and at the activation phosphory-

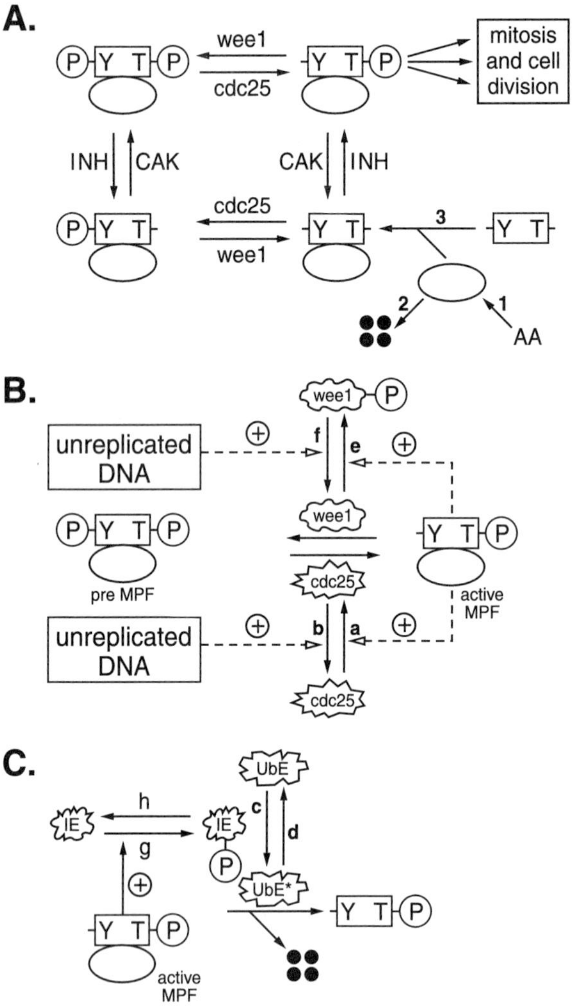

FIGURE 5.4 Schematic of the M-phase model. A. Shows the phosphorylation states of M-phase promoting factor (MPF). Ovals indicate cyclin subunits that are synthesized from amino acids (AA) (step 1). Squares with two phosphorylation sites (Y,T) indicate Cdc2 monomers. Degradation of free cyclin is noted by the small circles (step 2), Cdc2-cyclin dimerization is indicated by the combination of a square and oval (step 3). The phosphorylation state of Cdc2-cyclin dimers is indicated by the presence of phosphate groups at Tyr15 (Y) or Thr161 (T). The action of the kinase-phosphatase pairs Wee-1/Cdc25 and CAK/INH are indicated by the reaction arrows on the diagram. Phosphorylation of the Thr161 site (activation site) alone leads to the active complex, triggering mitosis and cell division. B. Positive feedback loops. Active MPF stimulates its own production by activating Cdc25 and inhibiting Wee-1. An external signal from unreplicated DNA to Cdc25 and Wee-1, providing a means to modulate the feedback loops. Note the phosphorylation states of Cdc25 and Wee-1 on the diagram. C. Negative feedback loops. Active MPF stimulates its own destruction by indirectly activating the ubiquitin-conjugating enzyme. An enzyme termed the intermediary enzyme (IE) adds a delay to the signal. Reprinted with permission from Reference 17.

lation site, a kinase named CAK and a phosphatase labeled INH. MPF is active, e.g., will initiate mitosis when the activation site is phosphorylated and the inhibitory site (actually two sites on the protein that are assumed to be processed simultaneously) is dephosphorylated. The dimer is ultimately separated by the degradation of cyclin via a ubiquitin-conjugating enzyme, which in turn is regulated by MPF itself. To account for a time lag between the increase in MPF and the degradation of cyclin, Novak and Tyson[17] introduce an intermediary enzyme in their model.

The schematic shown in Figure 5.4 was constructed based on experimental data from a number of sources that Novak and Tyson[17] compiled about M-phase control in Xenopus eggs and extracts. The next step in building the quantitative model is to write the differential equations that correspond to the model steps. The equations derived by Novak and Tyson[17] are shown in Figure 5.5. The differential equations describe the rate of change of the major variables in the system, including the different states of phosphorylation of the Cdc2-cyclin dimer. The equations can be written from inspection of Figure 5.4, using the law of mass action to derive the equations for the Cdc2-cyclin dimer and the cyclin monomer, and the Michaelis-Menten equation to model the kinetics of the regulatory enzymes: Cdc25, Wee-1, the intermediate enzyme, and the ubiquitin-conjugating enzyme.

The system of differential equations listed in Figure 5.5 can be solved numerically by using algorithms that include an integrator. These algorithms are implemented in a number of software packages including DBSolve, Mathematica, PhasePlane, and other mathematical modeling packages. DBSolve is a complete mathematical modeling workbench, freely available on the web, that includes capabilities to solve nonlinear algebraic equations, ordinary differential equations, bifurcation analysis, and fitting to experimental data.[18] A significant challenge for a model of this complexity is to assign values to the rate constants. If experimental data is available, it is used to set the rate constants. Otherwise, the rate constants are set in relative terms at magnitudes in accord with the assumptions that the investigator wants to test about the model. Choice of the rate constants impacts the algorithm used to solve the equations. If a system of equations and corresponding rate constants is characterized by some equations change that rapidly with time and others that change more slowly, the integrator must be able to dynamically adjust to the different time scales.

Comparison of model simulated results with experimental results is needed for model verification and to guide the process of setting rate constants. Often, families of curves are constructed using the model to understand the effect of initial conditions. Figure 5.6 shows the time course for active MPF as a function of initial cyclin concentration. Assumptions are made about the starting conditions: all Cdc2 is in the state of an unphosphorylated monomer, Wee-1 is in active form, Cdc25 is in the inactive form, and that no cyclin synthesis or degradation is simulated. The family of curves shows that there is a threshold for cyclin concentration around 30 to 40% of total Cdc2, associated with MPF activation. Increasing the initial cyclin concentration decreases the time lag between cyclin addition and the final level of active MPF (the asymptote of the time course). The time-course curves and a replotting of the 50% of maximal MPF concentration (equivalent in concept to the EC_{50} of a concentration response curve and shown as an inset in the figure) was in agreement with experimental results reported by Solomon et al.[19]

$$1.\quad \frac{d}{dt}[\bigcirc] = k_1[AA] - k_2[\bigcirc] - k_3[\bigcirc][\square]$$

$$2.\quad \frac{d}{dt}\left[\text{-}\overset{\square}{\bigcirc}\text{-}\right] = k_{INH}\left[\text{-}\overset{\square}{\bigcirc}\text{-P}\right] - (k_{wee}+k_{CAK}+k_2)\left[\text{-}\overset{\square}{\bigcirc}\text{-}\right] + k_{25}\left[\text{P-}\overset{\square}{\bigcirc}\text{-}\right] + k_3[\bigcirc][\square]$$

$$3.\quad \frac{d}{dt}\left[\text{P-}\overset{\square}{\bigcirc}\text{-}\right] = k_{wee}\left[\text{-}\overset{\square}{\bigcirc}\text{-}\right] - (k_{25}+k_{CAK}+k_2)\left[\text{P-}\overset{\square}{\bigcirc}\text{-}\right] + k_{INH}\left[\text{P-}\overset{\square}{\bigcirc}\text{-P}\right]$$

$$4.\quad \frac{d}{dt}\left[\text{P-}\overset{\square}{\bigcirc}\text{-P}\right] = k_{wee}\left[\text{-}\overset{\square}{\bigcirc}\text{-P}\right] - (k_{INH}+k_{25}+k_2)\left[\text{P-}\overset{\square}{\bigcirc}\text{-P}\right] + k_{CAK}\left[\text{P-}\overset{\square}{\bigcirc}\text{-}\right]$$

$$5.\quad \frac{d}{dt}\left[\text{-}\overset{\square}{\bigcirc}\text{-P}\right] = k_{CAK}\left[\text{-}\overset{\square}{\bigcirc}\text{-}\right] - (k_{INH}+k_{wee}+k_2)\left[\text{-}\overset{\square}{\bigcirc}\text{-P}\right] + k_{25}\left[\text{P-}\overset{\square}{\bigcirc}\text{-P}\right]$$

$$6.\quad \frac{d}{dt}[\text{-}\square\text{-}] = k_2\left(\left[\text{-}\overset{\square}{\bigcirc}\text{-}\right] + \left[\text{P-}\overset{\square}{\bigcirc}\text{-}\right] + \left[\text{P-}\overset{\square}{\bigcirc}\text{-P}\right] + \left[\text{-}\overset{\square}{\bigcirc}\text{-P}\right]\right) - k_3[\bigcirc][\square]$$

$$7.\quad \frac{d}{dt}[\text{cdc25-P}] = \frac{k_a\,[\text{active MPF}]\,([\text{total cdc25}] - [\text{cdc25-P}])}{K_a + [\text{total cdc25}] - [\text{cdc25-P}]} - \frac{k_b\,[\text{PPase}]\,[\text{cdc25-P}]}{K_b + [\text{cdc25-P}]}$$

$$8.\quad \frac{d}{dt}[\text{wee1-P}] = \frac{k_e\,[\text{active MPF}]\,([\text{total wee1}] - [\text{wee1-P}])}{K_e + [\text{total wee1}] - [\text{wee1-P}]} - \frac{k_f\,[\text{PPase}]\,[\text{wee1-P}]}{K_f + [\text{wee1-P}]}$$

$$9.\quad \frac{d}{dt}[\text{IE-P}] = \frac{k_g\,[\text{active MPF}]\,([\text{total IE}] - [\text{IE-P}])}{K_g + [\text{total IE}] - [\text{IE-P}]} - \frac{k_h\,[\text{PPase}]\,[\text{IE-P}]}{K_h + [\text{IE-P}]}$$

$$10.\quad \frac{d}{dt}[\text{UbE*}] = \frac{k_c\,[\text{IE-P}]\,([\text{total UbE}] - [\text{UbE*}])}{K_c + [\text{total UbE}] - [\text{UbE*}]} - \frac{k_d\,[\text{anti-IE}]\,[\text{UbE*}]}{K_d + [\text{UbE*}]}$$

$$11.\quad k_{25} = V_{25}'\,([\text{total cdc25}] - [\text{cdc25-P}]) + V_{25}''\,[\text{cdc25-P}]$$

$$12.\quad k_{wee} = V_{wee}'\,[\text{wee1-P}] + V_{wee}''\,([\text{total wee1}] - [\text{wee1-P}])$$

$$13.\quad k_2 = V_2'\,([\text{total UbE}] - [\text{UbE*}]) + V_2''\,[\text{UbE*}]$$

FIGURE 5.5 The system of differential equations for the M-phase model. The circles and squares indicate cyclin and Cdc2, respectively. Corresponding to Figure 5.4, the cyclin-Cdc2 dimers and associated phosphorylation states are indicated by symbols in the equations. The small letters on the rate constants correspond to arrows on the reaction arrows in Figure 5.4. Reprinted with permission from Reference 17.

The relationship between total cyclin concentration and active MPF is complex as a result of the threshold. The equilibrium relationship between these variables is derived from the model and shown in Figure 5.6B. The arrows indicate the trajectory of the state of active MPF for a given starting concentration of cyclin. After some defined

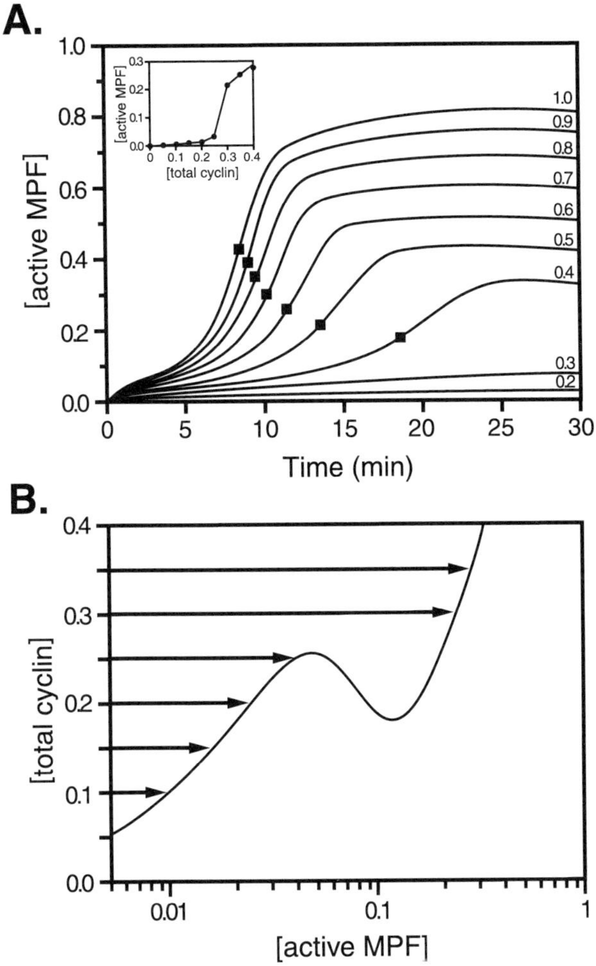

FIGURE 5.6 Cyclin-induced activation of MPF. A. Shows the concentration of active MPF simulated by the mathematical model. The number at the far right on each curve indicates the amount of cyclin added. The black squares mark the lag point at 50% of final MPF activity. The inset shows final MPF activity as a function of cyclin level. Abscissa: Time in minutes. Ordinate: Concentration of active MPF. B. Total cyclin as a function of active MPF concentration. As total cyclin is increased, there is a threshold for activation of MPF. Arrows indicate the trajectory of the state of active MPF. Abscissa: Active MPF concentration. Ordinate: Total cyclin concentration. Reprinted with permission from Reference 17.

period of time the equilibrium concentration of active MPF is reached. For the portion of the curve below the threshold (bend in the curve), small increases in cyclin yield only small increases in MPF. Above the bend in the curve, increases in cyclin yield rapid increases in MPF as a result of the positive feedback loop in the model. In other parts of their paper, Novak and Tyson[17] present results of biochemical interventions on the cyclin/MPF relationship and the oscillatory dynamics of their model system.

The approach outlined by Novak and Tyson[17] is applicable to modeling of signal transduction and cellular regulation processes. The mathematical framework of expressing the schematic of the process as a system of differential equations, derived using enzyme kinetics, law of conservation of mass, and law of mass action is a well-established process in biochemistry. Obtaining experimental data to determine the values of rate constants and to test the model is critical to the process. Integrating the information from proteomics and gene expression experiments in a form that will facilitate writing systems of equations is a challenge.

5.6 CASE STUDY 2: VISUALIZATION AND CLUSTERING OF DATA FROM BIOINFORMATICS AND CHEMINFORMATICS

The second case study involves correlating structure-activity relationship data from the National Cancer Institute database of activities of compounds against different cancer cell lines with information about biochemical characteristics of the cell lines.[20] It is an excellent example of the integration of information from bioinformatics and cheminformatics.

The structure of the problem is shown in Figure 5.7. There are three matrices of information available from the databases: a matrix of structure-activity relationship (SAR) data which consists of measured activities of compounds against each of the cell lines, a matrix of chemical descriptors for the compounds, and a target matrix that defines whether a cell line contains specific molecular characteristics. These characteristics are potential sites of interaction in a regulatory network (receptors, oncogenes, tumor-suppressor genes, metabolic enzymes, and other intracellular signaling molecules). The SAR data is obtained from testing the compounds in a bioassay. The chemical descriptors are calculated from the 2D or 3D structures. The target matrix is constructed using knowledge about the cell lines.

When the correlations between pairs of descriptors are examined, there are several visualization methods that can be used. Figure 5.8 shows a gray-scale correlation matrix for a group of compounds compared to their presumed biochemical targets of action. The intensity of the cell indicates the strength of the correlation between the compound and the specific enzyme target. The correlation patterns can be used to assess both the specificity of the compounds or to formulate hypotheses about the mode of action of the compounds.

A graphical correlation matrix is an alternative to using single cells (r statistic or gray-scale indicator) for each correlation. Each cell is a graphical representation of the data that are used in the correlation coefficient calculation. In Figure 5.9, the binding affinities of a set of compounds to different cloned human α_1-adrenoceptor cells and to canine and human prostate membranes is shown. The width of the ellipse indicates strength of the correlation, a narrower ellipse indicates a stronger correlation. The plot allows an analyst to assess the strength of different correlations, determine whether the correlations are specific for chemical classes, and to identify which compounds are most influential in the correlation.

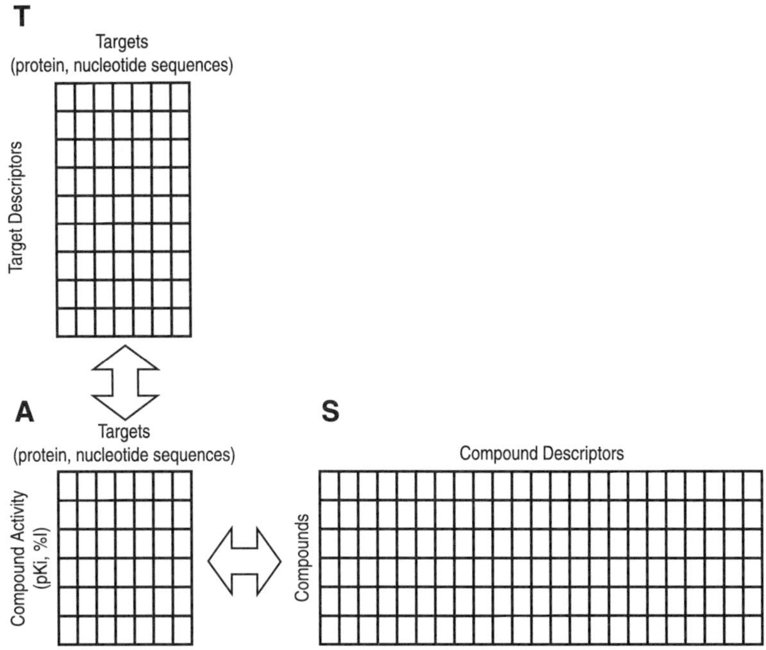

FIGURE 5.7 Schematic of the 3-way problem. Sequence and compound structure-activity relationships are represented by three matrices. The upper left matrix (T) consists of descriptors for protein or nucleotide sequences. Columns in the matrix represent each sequence of interest, rows will be equal to the number of descriptors calculated for each sequence. The lower left matrix (A) is observed activity for a set of compounds. There is a row for each compound in the set. Columns consist of measurements for each of the proteins or receptors in the analysis (e.g., correspond to columns in the matrix above). The matrix at the right (S) consists of compound descriptors. There is a row for each compound in the set and a column for each descriptor.

Information from the three matrices shown in Figure 5.7 can be combined using matrix mathematics and clustering methods. Figure 5.10 shows four different graphical correlation matrices derived from the NCI data.[20] Clustering is done for the rows, or columns, or both to group the corresponding data. Figure 5.10A is a matrix of compound activity (rows) by targets where the activity vectors were randomized. The lack of any pattern indicates that the clustering does not create spurious patterns in the data. Figure 5.10B is a cross-correlation of the target matrix where no compound data was used in the analysis. Areas of high correlation indicate concomitant expression of several molecular characteristics in the cell lines. Clusters of molecular characteristics related to Mdr-1, heat shock proteins (hsp), and p53 functions are identified from this plot. Figure 5.10C relates the target and SAR matrices. In this plot, the activity profiles for the 3989 compounds are used to calculate the correlations. The same protein families are identified in this plot, indicating that for the compounds in the dataset, the target classification has a strong association with

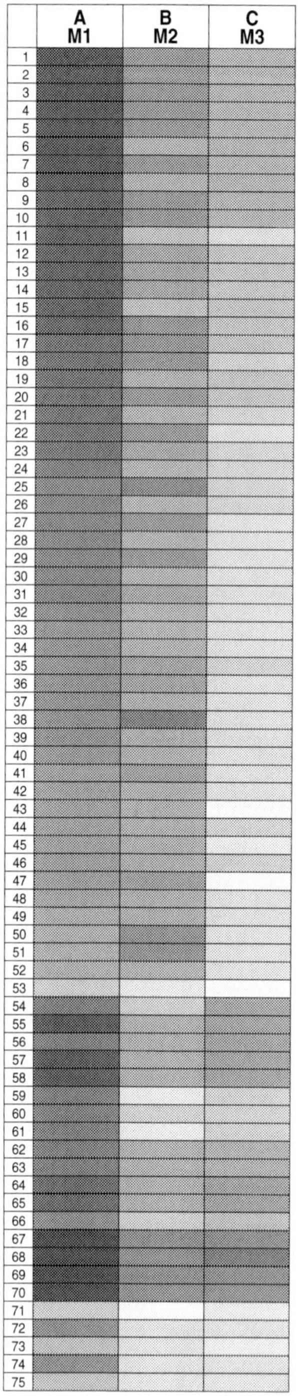

FIGURE 5.8 Visual correlation matrix. This matrix shows pairwise correlation coefficients for the affinity of a set of 75 compounds tested against three mutants of an HIV protease. The strength of the correlation is indicated by the intensity of the square. Compounds that are relatively more selective for one mutant show a discontinuous pattern while compounds that are nonselective show a uniform pattern across all three squares. For large amount of data, this method of visualization method is useful for identifying patterns that are of interest for further structure-activity relationship modeling. Data from Eric Furfine, Glaxo Wellcome.

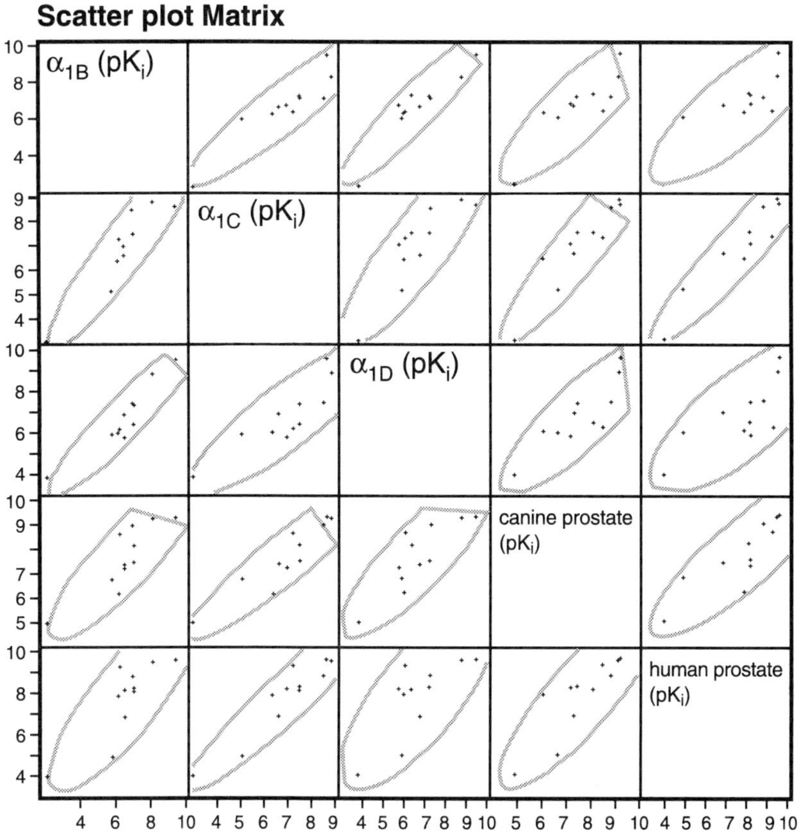

Variable	α1B (pKi)	α1C (pKi)	α1D (pKi)	canine prostate (pKi)	human prostate (pKi)
α1B (pki)	1.00	0.92	0.93	0.86	0.83
α1C (pki)	0.92	1.00	0.87	0.93	0.94
α1D (pki)	0.93	0.87	1.00	0.85	0.75
canine prostate (pKi)	0.86	0.93	0.85	1.00	0.85
human prostate (pKi)	0.83	0.94	0.75	0.85	1.00

FIGURE 5.9 Correlating SAR activity patterns. Each cell is a graphical representation of the data that are used in the correlation coefficient calculation. The units of all axes are $-\log(K_i)$, that is, pK_i, for binding to various cloned human α_1-adrenoceptor cells and to canine and human prostate membranes. The width of the ellipse indicates strength of the correlation, a narrower ellipse indicates a stronger correlation. Reprinted with permission from Reference 34.

the pattern of activity data. Figure 5.10D is the product of all three matrices. This plot adds information about chemical substructure to the plot.

Use of clustering and visualization methods as illustrated in this case study is an effective means of generating hypotheses and models about the molecular basis of cancer. Statistical models can be constructed to relate the data matrices in order to make predictions about other compounds or targets. For example, if the target expression pattern for a new cell line were known, the model would predict activity, which

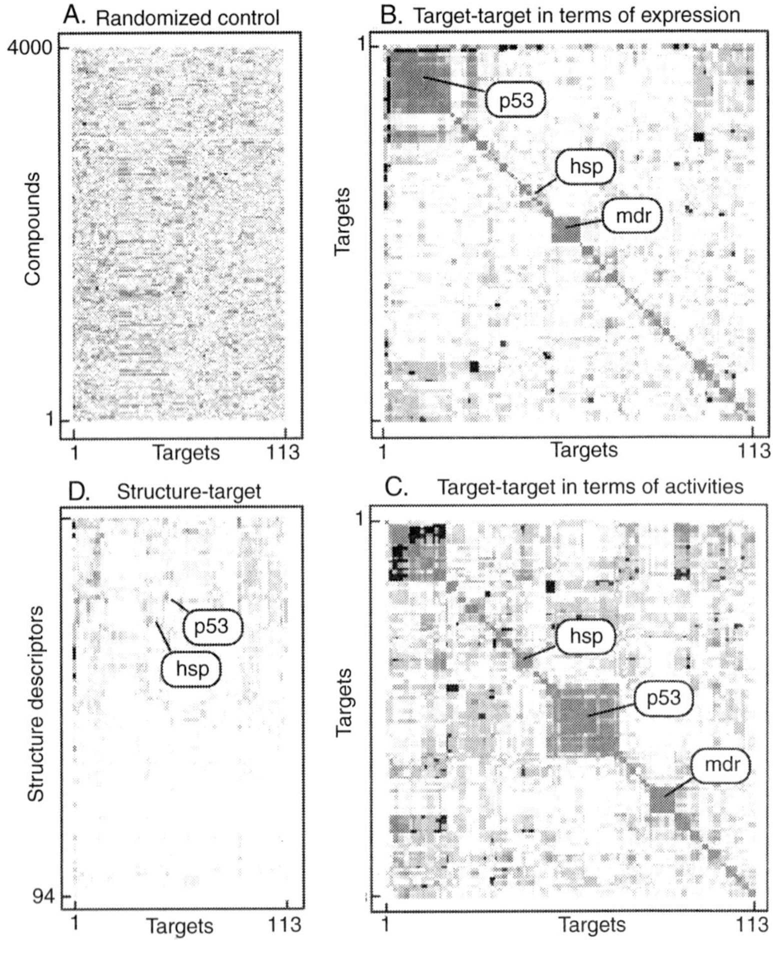

FIGURE 5.10 Clustered correlation matrices. Matrix designations correspond to the labels in Figure 5.7. In all cases, the matrix calculations are done as follows: 1) each row of the matrix is normalized by its mean and standard deviation, 2) products are a result of matrix multiplication, ′ indicates matrix transposition, 3) each entry is divided by n-1 to produce a Pearson correlation coefficient ($-1 \leq r \leq 1$), 4) clustering is done on the rows, columns, or both. A. This matrix is similar in format to Figure 5.8. Rows are compounds and columns are targets. The intensity of each cell is representative of the activity of the compound against a target. In this figure, the activity vectors were randomly permuted, however, to demonstrate that there is no pattern in the data produced by the clustering process itself. B. Cross-correlation of the target matrix. Mathematically, this corresponds to the product of $T \cdot T'$. Targets with the most similar expression pattern appear side by side after clustering. The diagonal is a consequence of the $r = 1$ correlation of each target with itself. C. In this plot, the targets are characterized by the structure-activity relationships derived from the biological activity data. This is also a cross-correlation matrix with a format similar to B. The matrix corresponds to $(A \cdot T') \cdot (A \cdot T')$. D. Product of all three matrices, $S' \cdot (A \cdot T')$. This plot adds information about chemical substructure to the matrix. Reprinted with permission from Reference 20.

would in turn be tested with experimental data. For the chemistry matrix, calculation of descriptors for untested compounds would allow prediction of the activity pattern for the different cell lines, again allowing comparison to experimental data.

5.7 CASE STUDY 3: CLASSIFICATION OF G PROTEIN-COUPLED RECEPTORS USING HIDDEN MARKOV MODELS

Hidden Markov models are an important class of statistical models that are used in bioinformatics. These models are a method for identifying patterns in nucleic acid or protein sequence data, gene finding (locating exons, promoter regions, enhancers, splice sites), and searching sequence databases for entries that correspond to the profile of a target set of sequences (multiple sequence alignments, classification of groups of sequences, database mining). Sequence searching is a challenging problem from the perspective that the search must deal with the fuzziness of biological sequences. Unlike a substructure search in chemistry, which is often an exact match to a target, families of biological sequences often have insertions, deletions, and a relatively low level of homology. Proteins having similar function may have less than 20% homology at the sequence level. However, it is important to recognize that the sequences are similar, even if the homology is low, in order to cluster families of proteins and to identify subtle matches between a target protein and entries in a database.

Structurally, a hidden Markov model is composed of states and transitions. A schematic of a hidden Markov model of a nucleic acid sequence is shown in Figure 5.11. Probabilities are assigned to each transition between states (arrows between the

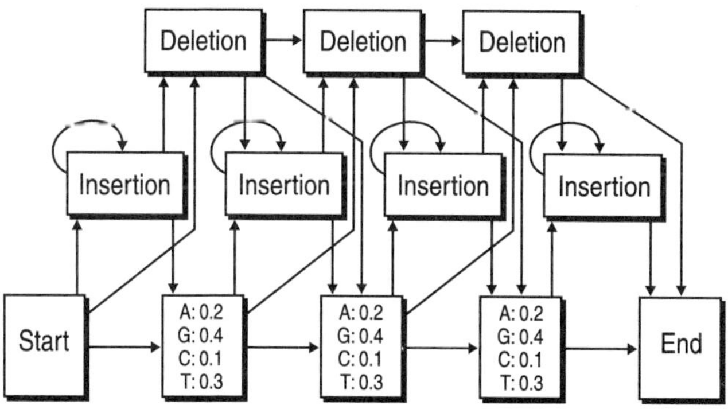

FIGURE 5.11 Schematic diagram of a hidden Markov model. The states, or boxes at the bottom of the diagram correspond to each residue position in the sequence. The probability for each nucleotide appearing at a position is given in the state box. The arrows represent the probability of moving between the states. The middle row of boxes handles insertions. The cycle arrow indicates that the insertion is of a variable length. The top row handles deletions and allows a transition from one state to a subsequent state. Note that all transitions can be modeled by this diagram. From each main state, progression is possible to the adjacent state, an insertion state, or a deletion state.

boxes that represent the states). Recursive transitions are allowed and are the mechanism used to represent insertions of variable length (depicted by the circular arrow above a state box). Each state box contains a list of possible values (for the DNA sequence, GCAT) and the probabilities associated with each value. The complete model, with all probabilities assigned, is in fact a pattern that can be compared to any arbitrary sequence. Probabilities are assigned based on prior knowledge, a specific motif to be matched, or a multiple sequence alignment (e.g., the motif to be matched is derived from a collection of sequences). Each sequence is evaluated in the context of the model and an overall score is assigned by multiplying the probabilities: the product of observed state and transition probabilities. Higher values of this score correspond to a better fit to the pattern defined by the model. In practice, logarithms of probabilities are calculated so that the overall score can be computed as the sum of log-odds. To model deletions in the sequence, an additional state variable is used that permits bypassing states. In Figure 5.11, each residue position is represented by the lower row of state boxes. The middle row handles insertions while the top row is used to model deletions. Note that any pattern of transitions can be modeled or any combination of insertions and deletions and transitions between states.

A hidden Markov model can be trained from a collection of sequences in a process similar to the training of a supervised neural network. Dynamic programming is often used to train the model using unaligned sequences in an iterative fashion similar to nonlinear regression with the desireable result that a multiple sequence alignment is created as a byproduct of the optimization phase. A model with initial estimates of probabilities is used as a starting point. The collection of sequences is used to refine the probabilities in a systematic way. The new probabilities, in turn, may identify a different optimal alignment. The process is repeated, trading off between alignment and probability coefficient optimization until the alignment no longer changes (e.g., convergence). After the model is created, new sequences, are used as the input to the model, aligned with the training sequences and scores calculated. As with nonlinear regression, local minima are a problem. The alignment, the probabilities, or both may be suboptimal. In practice, methods that use Markov chain models for gene finding try to minimize the chances for local minima by using dynamic programming to work on subproblems (e.g., searching for donor and acceptor sites) and by integrating the model with other approaches, e.g., decision trees.

In this case study, hidden Markov models are used to classify fragments of sequences. Databases are available of expressed sequence tags (EST) and cDNA from genomic sequencing. The problem is to use pattern recognition methods to classify the fragments into likely protein families (7TM, kinases). Classification of orphan receptors and clustering of sequences and sequence fragments into subfamilies are extensions of this type of problem.

Baldi and Brunak[21] utilized a hidden Markov model to simulate the problem of identifying G protein-coupled receptors (GPCRs) from protein fragments. Their simulations examined the effects of sequencing noise and sequence length on the identification problem. The simulation was designed to indicate that a sequence was a GPCR; other classification problems involve clustering sequences or fragments into families. The Markov model was trained using 142 GPCR sequences from the PROSITE database. The average length of the training sequences was 430 amino acids. After

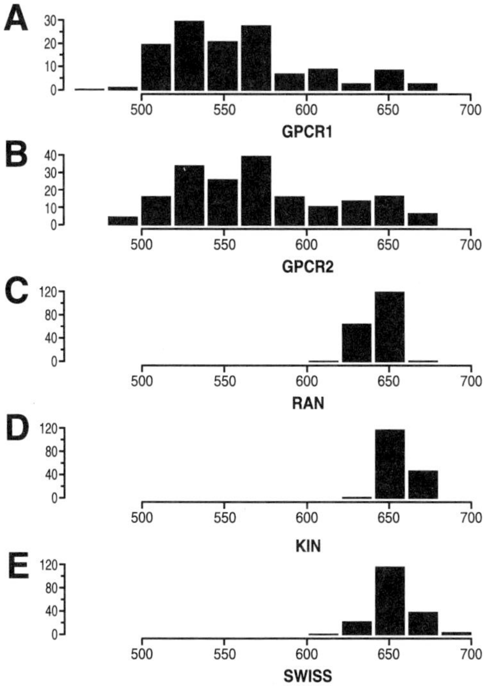

FIGURE 5.12 Recognition of G protein-coupled receptors (GPCRs) using a hidden Markov model (HMM). The histograms are of scores from the HMM for five different datasets. The fragments are all of length 150. A. 142 random sequence fragments from the training set (all GPCRs). B. 200 sequence fragments taken from a GPCR database. C. 200 sequence fragments taken from randomly generated sequences with a similar average composition to the training set. D. 200 sequence fragments taken from a kinase database. E. 200 sequence fragments taken from the SWISS-PROT database. Abscissa: Score from the HMM. Ordinate: Count of number of sequences. Reprinted with permission from Reference 21.

training, any sequence or sequence fragment evaluated by the model is scored by calculation of the negative log-likelihood of a corresponding path through the model.

Figure 5.12 shows the results of using the hidden Markov model developed by Baldi and Brunak[21] to classify five different sets of sequences. The histogram shows counts of fragments having specific negative log-likelihood scores from 500 to 700. For these simulations, sequences of length 150 amino acids were used with no sequencing noise added. Figure 5.12A is constructed from random fragments taken from the training set, Figure 5.12B is calculated from 200 fragments taken from a GPCR database, Figure 5.12C is calculated by generating random sequences of length 150 with average composition equal to the GPCR sequences, Figure 5.12D is derived from a database of kinase sequences, and Figure 5.12E is constructed by taking random sequences from the SWISS-PROT database. The distributions are markedly different. In particular, it appears that setting a threshold score of 625 would be appropriate for classifying GPCRs. As fragment length is decreased and sequencing noise added, classification becomes more difficult. Figure 5.13 shows

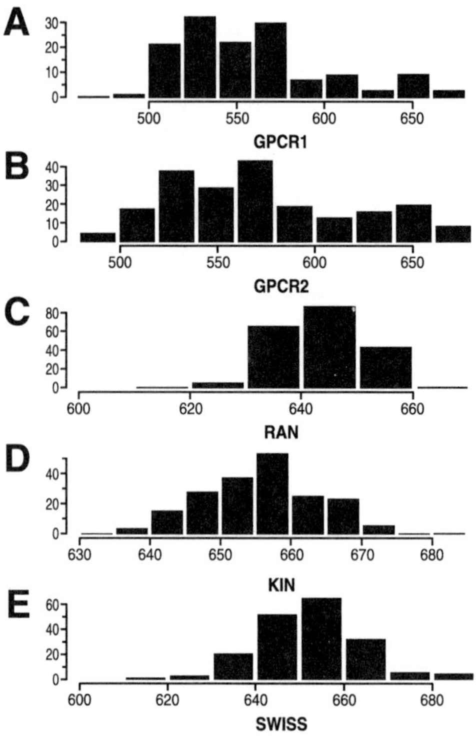

FIGURE 5.13 Adding sequencing noise and reducing fragment length: effect on recognition of GPCRs. Figure is similar to Figure 5.12 with the exception that sequence fragments of length 50 are used and 10% sequencing noise is added. Reprinted with permission from Reference 21.

frequency distributions for the same sequence samples, but for fragments of length 50 instead of 150 and 10% sequencing noise added. The distributions from the non-GPCR sequences are wider and have more overlap with the GPCR distributions. A systematic analysis of sequence length and noise is shown in Figure 5.14. By examination of this plot, the probability of discrimination between target (GPCR) sequences and nontarget sequences in SWISS-PROT is estimated. Baldi and Brunak[21] used these simulations to estimate true and false positive rates and draw receiver operating characteristic curves. Longer fragments that include more of the conserved regions of the protein are easier to discriminate. Short fragment lengths and high levels of noise make it more difficult to identify the GPCR sequences.

5.8 CASE STUDY 4: PREDICTION OF FUNCTIONAL REGIONS OF PROTEINS FROM THEIR CODING SEQUENCES USING A NEURAL NETWORK

The nucleotide sequences of human genes are a starting point to derive biological information. Using the tools of genomics, promoters, DNA binding sites, mRNA

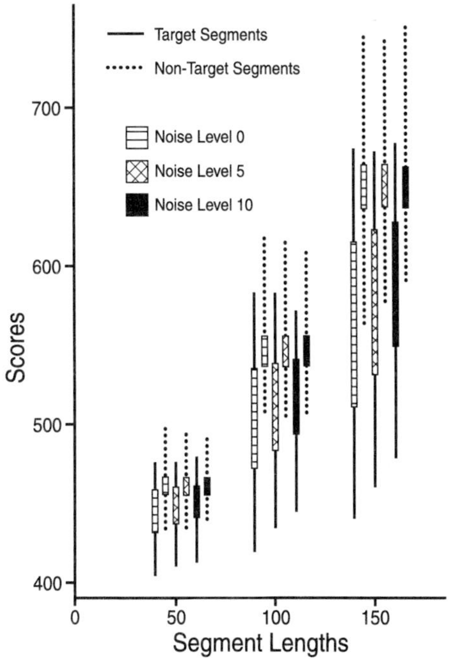

FIGURE 5.14 Summary of the HMM/GPCR work. This figure shows the impact of sequencing noise and sequence fragment length on classification using the HMM filter to recognize GPCRs in SWISS-PROT. Target sequence scores are shown with solid lines. The standard deviations are depicted with the larger rectangles and are coded by noise levels. The ranges are shown as the tails on the boxes. Nontarget sequence fragments are shown using dotted lines. Abscissa: Sequence fragment length. Ordinate: Score from the HMM. Reprinted with permission from Reference 21.

splice sites and coding, regions of genes are identified. Ultimately, the correlation between the coding regions of sequences (exons) and their functional proteins is necessary to understand the genetic basis of cellular processes and disease. This case study illustrates the use of an unsupervised neural network to correlate coding sequence with function in families of GPCRs.[22] The case study was chosen to illustrate the application of an unsupervised neural network as a computational model and to demonstrate the use of information from a bioinformatics database. As with several of the case studies in this chapter, the mathematical model is used for a pattern classification or clustering problem.

From a GPCR database (http://swift.embl-heidelberg.de/7tm/htmlsGPCR.html), sequences for 145 receptors were selected and grouped into families according to their activating ligand. Four families were represented in this set: adrenergic, muscarinic, serotonin, and opsin. Each of the sequences has a corresponding entry in the protein database, SWISS-PROT, which contains a putative topological model. The input to the neural network is the coding sequence of the receptor.

Unlike neural networks that involve supervised learning (e.g., backpropagation), unsupervised neural networks do not use a training set to relate an input vector to an output. Instead, the network self-organizes to patterns in a data set, retaining the most statistically significant features of the data. In the unsupervised neural network, only one output neuron is active at a given time. In a similar manner to recursive partitioning and classification and regression trees, self-organizing neural networks are used to identify features that are important for classifying data and therefore are suited to the problems of identification of molecular features that are important for biological activity from high-throughput screening data[23] and, in this example, correlating coding sequence with function. Other approaches for assigning function to sequences are based on hidden Markov models, Bayesian statistics,[24] and Gibbs sampling methods.[25]

For the case study, the neural network can be considered a filter that identifies segments in coding sequences that correlate with function. If a coding sequence is identified as important, the segment is marked on the full length protein sequence. The transmembrane topology is shown as annotated in the protein database, SWISS-PROT. In the process of extracting features from the sequences, windows of different lengths of nucleotides (9, 12, 15, 18, and 21) are used to scan the sequence.

The test of this approach is to determine whether the neural network identifies functionally important features. The algorithm used by the authors requires that four out of five windows signal activated neurons to declare a region of the sequence a significant pattern. If the windows do overlap, the pattern of amino acids covers the range of all of the windows; e.g., the pattern will be as long as the union of the windows. Figure 5.15 shows the results of applying the filter to a single sequence, the human β_2 receptor. The amino acid sequence is at the top and the 7-transmembrane regions are marked underneath. Significant patterns are identified by the filter in the N-terminus, 1st and 7th transmembrane regions, 3rd intracellular loop, and the C-tail. Assessing the biological relevance of these patterns is done by comparing the patterns observed in different sequences and by designing experimental work such as building chimeric receptors. The latter is an important application of mathematical models: the simulation models and the pattern recognition methods are a means of planning experimental work to test hypotheses generated by the models.

In their study, the authors compared the patterns observed in the human β_2 and β_3 receptors. They found that for β_2, there were significant signals in the i2 and i3 (intracellular) regions, first and seventh transmembrane (TM) regions, and e2 (extracellular) not observed in β_3, while β_3 demonstrates signals in the 5th TM region not seen in β_2. The authors used data from chimeric receptor studies of desensitization[26] to compare with the patterns recognized by their filter. Unlike the β_2 receptor, the β_3 receptor is resistant to short-term agonist-promoted desensitization. Chimeric receptor studies demonstrated that i2, i3, and C-tail substitutions were all essential for agonist-promoted desensitization, demonstrating that these regions of the protein determine a regulatory phenotype for desensitization. The authors concluded that the two sequences contain intrinsic features related to functional specificity for desensitization, activation, and sequestration. In addition to the work with the β receptors, the authors analyzed G protein-coupling specificity of the rat m3 muscarinic receptor, again comparing the results of the filter to experimental data from chimeric receptors.[22]

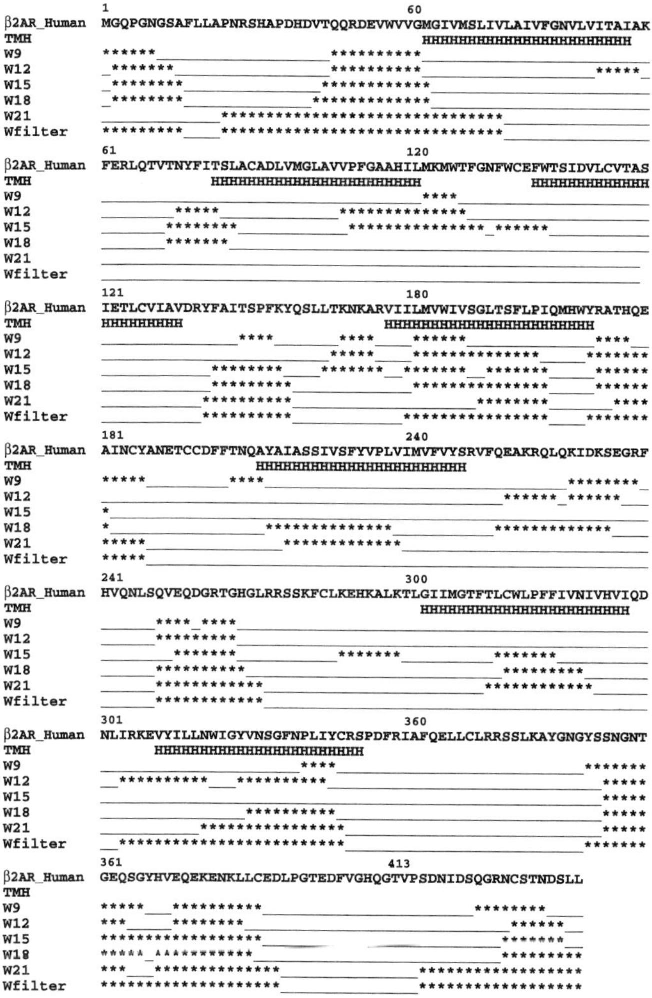

FIGURE 5.15 Analysis of signal (β_2) from neural network. The top row of the figure is the amino acid sequence. The second row indicates the transmembrane regions. Rows three to seven are the output from windows of different lengths of nucleotides as indicated in the text. The bottom row is the consensus output from the filter. Four out of five windows must have an active signal for the consensus to be active. Reprinted with permission from Reference 22.

In addition to work with single sequences, a classifier such as this filter can be applied to multiple sequences to determine if there are patterns or motifs that are characteristic of subfamilies or of all receptors in a family. Arrigo et al.[22] used their filter to identify 904 segments from 145 coding sequences of 5 general subfamilies of 7TM receptors (muscarinic, adrenergic, serotonin, photoreceptor, and hormone). The protein sequences in the subfamily are aligned and the frequency of occurrence of the segments is compared to the average frequency of occurrence of all segments plus one standard deviation unit. This allows construction of a plot that shows, for

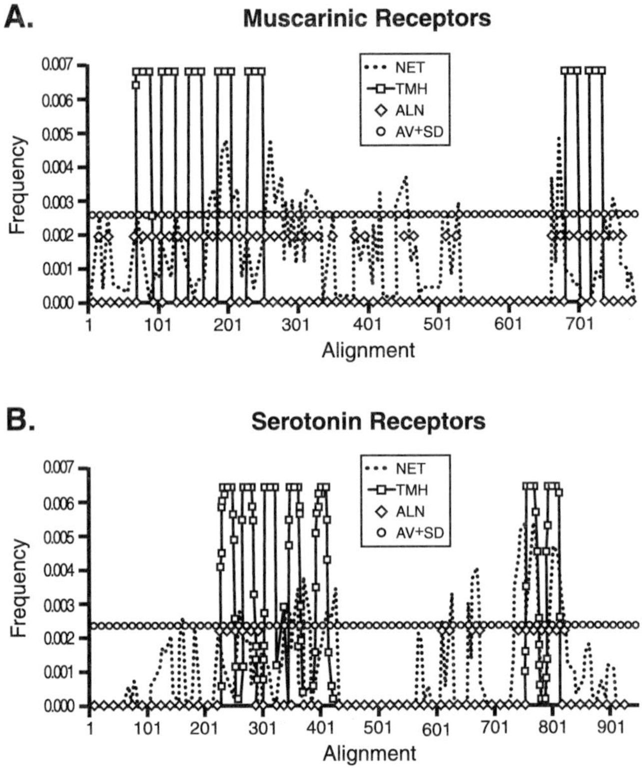

FIGURE 5.16 Classification of functional regions of GPCRs using a neural network. These histograms show the frequency of coding sequences extracted by the filter for two families of GPCRs. Plots such as these define a characteristic signal for a subfamily of proteins. The curve labeled NET is the signal frequency from the filter. TMH is the transmembrane position of the putative transmembrane regions of the receptors as a function of alignment. ALN is the residue frequency in the alignment. AV + SD is the average frequency of occurrence of all the signals plus one standard deviation. Signals greater than AV + SD are indicative of a robust, strong signal. The plot also allows a comparison of the filter signal with the conserved regions of the protein. A. Muscarinic receptors. B. Serotonin receptors. Abscissa: Position in the multiple sequence (protein) alignment. Ordinate: Frequency. Reprinted with permission from Reference 22.

each residue position, the signal compared to the average (Figure 5.16). The transmembrane topology is shown on the same plot. Figure 5.16 compares the signals for the muscarinic and serotonin subfamilies. Note the strong signal in transmembrane region 4 for both families. Differences in signal are seen in the 6th and 7th transmembrane regions; note the stronger signal in the serotonin subfamily for these regions. The authors found literature precedent for functional significance of the 6th transmembrane region (e.g., site-directed mutagenesis or chimeric receptor experiments). Their results are summarized for all of the receptor subfamilies in Table 5.2.

From the perspective of computational modeling, this case study is notable in that it uses a pattern recognition algorithm to identify features of functional impor-

TABLE 5.2
Significant Domains for GPCRs Extracted Using the Neural Network Filter Compared to Functional Domains with Literature Precedent

Subfamily	Extracted Domains	Signal Composition	Functional Domains
Adrenergic	i1	Q, N, T, F, Y, H, I, L	—
	II TMH	S, V, N, M, T	—
	IV TMH	W, I, M, P, H, S, F, L, V	*
	e2	Q, D, A, P, E, G	—
	V TMH	F, Y, P, S	*
	i3	K, H, R, E, Q, T, S, G	*
	VI TMH	P, M, W, V, C, I, L	*
	VII TMH	N, Y, F, I, C, S, D, P	*
Muscarinic	i2	R, K, T, Y	*
	IV TMH	W, A, F, L, I	*
	e2	G, R, V, F, Y, T, K	—
	i3	E, K, R, Q, S, A	*
	Ct	C	*
Serotonin	Nt	N, S, T, D, E, W, P, H, A, M	—
	i1	H, N, Y, K, Q, P, A	—
	i2	R, K, T, A, N	—
	IV TMH	P, W, I, F, G, L	—
	e2	D, C, E, K, Q	—
	V TMH	Y, I, L, V, F, W, T	*
	i3	H, K, R, Q, E, P, S	—
	VI TMH	F, K, M, R, I	*
	VII TMH	Y, N, W, F, I, L	—
	Ct	R, F, A, Y, Q, K	—
Photoreceptors	Nt	N, R, D, T, P, V, M	*
	I TMH	V, Y, T, F, W, I	—
	i1	K, H, R, L, N, T, P	*
	V TMH	H, P, C, F, L, I	—
	i3	Q, E, K, R, H	*
Hormones	Nt	N, E, R, Q, L, K, H	*
	IV TMH	W, M, G, V, F, A, L, I, P	—
	i3	K, D, R, T, A, N, S	*

Note: For each subfamily, the extracted domains and signal composition are determined by the neural network filter. Signals are all above the average frequency plus one standard deviation (AV + SD) as indicated in Figure 5.16. Nt is the amino terminus (extracytoplasmic), Ct is the carboxy terminus (intracytoplasmic). The intercytoplasmic loops are labeled i1–i3, the extracytoplasmic loops are labeled e1–e3. Transmembrane helices start at the N terminus and are numbered I–VII. Residues in the second column are those whose frequency of occurrence is higher than that in the receptor subfamilies. The functional domains are culled from the literature and represent evidence by hybrid construction, site directed, deletion, or random saturation mutagenesis and test peptides. See Ref. 22 for more details. Reprinted with permission from Ref. 22.

tance from the amino acid sequence. The significance of these features is assessed by searching for literature precedence through work with chimeric receptors or site-directed mutagenesis. In practice, pattern recognition methods are often used to design experiments, for example to suggest experiments for the construction of chimeric receptors or assays to measure a functional response.

5.9 CASE STUDY 5: ANALYSIS OF STRUCTURE-ACTIVITY RELATIONSHIP DATA USING RECURSIVE PARTITIONING

In drug discovery research, determining the relationships between chemical structure and a variety of measured responses (binding affinity, activity in functional assays, *in vivo* activity, metabolism, and toxicity) is the basis for optimizing a molecule for use as a drug. Models used for structure-activity relationship analysis include simple intuitive pattern recognition, e.g., looking at chemical structures and associated response data, statistical regression techniques (linear regression using descriptors calculated from the chemical structure), and other statistical classification models. Linear regression methods are often used for cases where the compounds consist of a small (less than 100) set of similar compounds. For large heterogeneous sets of thousands of compounds, recursive partitioning has been demonstrated to be an effective model to identify structure-activity relationships in data.[6,27]

The principle of recursive partitioning analysis is illustrated in Figure 5.17. Descriptors are derived from the chemical structure. Literature examples of descrip-

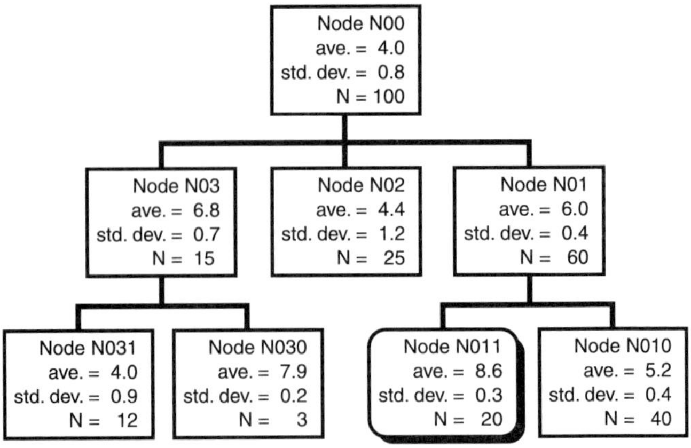

FIGURE 5.17 Simple schematic of recursive partitioning. Recursive partitioning groups the members of a collection of elements into nodes using a statistical algorithm. The entire collection is represented by the parent (top) node. Splitting rules based on descriptors or predictors calculated for the elements of the collection are used to form the branches of the tree. The tree can consist of entirely binary splits or multiway splits at each branch. Terminal nodes consist of a subset of the original collection that share a set of features (values of the descriptors) and cannot be split into subgroups, either because a significant splitting rule does not exist or there is insufficient statistical power in the sample to split the group further.

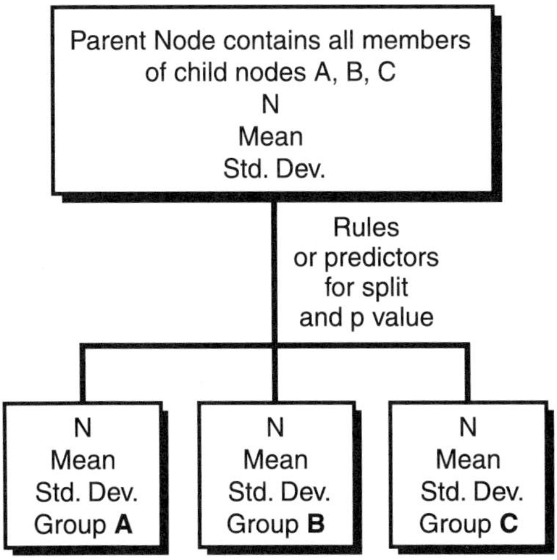

FIGURE 5.18 Single branch of a recursive partitioning tree. Each node or collection of observations is listed with the number of observations, mean, and standard deviation. Members of the group in a node cannot be split further on the basis of the set of predictors in the analysis and the corresponding value of the dependent variable. At each node, the predictor used to form the split is identified along with the p value associated with the split. Smaller values of the p value indicate more statistical significance for the split.

tors used for recursive partitioning analysis include atom pairs and topological torsions derived from the 2D chemical structure[27] and features derived from 3D conformations,[6] although technically, any descriptor can be used in the analysis including classification by substructure. The response variable is usually continuous, obtained from a biochemical assay. The analysis involves using the descriptors as splitting rules to partition the entire dataset into groups where the members of each group all have a feature identified by a rule and all have values of the response variable that are not statistically different from each other using a t-test. The results are expressed as a tree where the nodes or leaves of the tree contain the groups of compounds and the split above a pair or set of nodes identifies the splitting rule. Figure 5.18 shows a single branch of a tree. Each node of the tree or dendrogram lists the mean, standard deviation, and number of compounds contained in that node. The horizontal lines represent the rule on which the previous node was split. Compounds in the right-hand side node satisfy the rule; compounds in the left-hand side do not. The p value from the hypothesis test for this rule is noted on the tree. Finding the structural features associated with the most active molecules becomes a process of identifying the node(s) that has the highest mean activity and then identifying the structural features of the molecules that comprise that node by working backwards up the tree.

Recursive partitioning algorithms and software are available for both binary and multiway splitting of data at each branch of the tree. For Figure 5.17, all of the potential rules are tested to identify the single rule that best partitions the data into two groups, the resulting rule having the smallest p value for the t-test. This

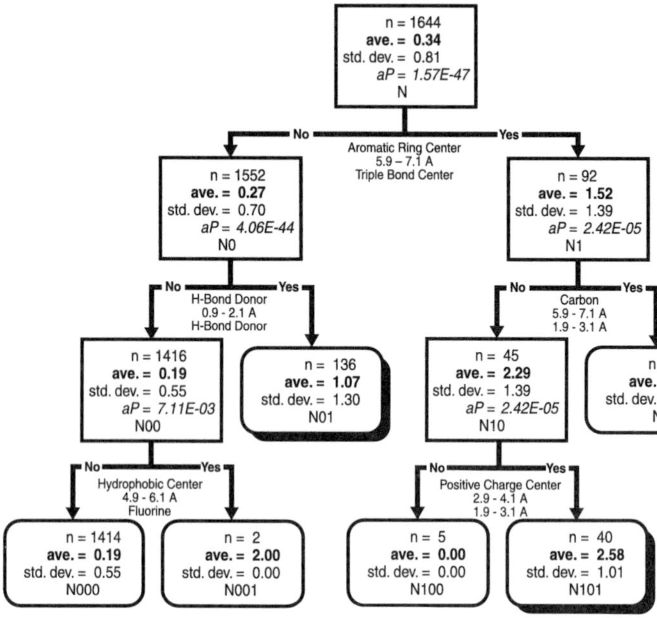

FIGURE 5.19 Tree based on 3D pharmacophore model. Recursive partitioning tree for monoamine oxidase inhibitors based on 3D pharmacophores. For each node, n is the number of compounds, ave is the average potency of the compounds, std dev is the standard deviation, and aP is the Bonferroni-adjusted p value for the t-test associated with the split. Active nodes are indicated by shading (N01 and N101). Two-point descriptors (pharmacophores) are indicated as two chemical features with the distance range between them, three-point pharmacophores are indicated as a chemical feature and two distance ranges. Reprinted with permission from Reference 6.

process is repeated for each of the resulting nodes so that other statistically significant positions and interactions can be detected. The process ends when either the number of compounds remaining in a subset is too small for additional splits to occur, or none of the remaining rules have enough statistical power to split the data further.

An example of using recursive-partitioning analysis to identify pharmacophores is given by Chen et al.[6] In this case study, a set of 1644 monoamine oxidase inhibitors was analyzed based on the chemical features listed in Table 5.1. The binary tree that results from this analysis partitions the data into groups of compounds that have different combinations of pharmacophoric elements. The first major split of the data isolates a subset of 92 compounds from the remainder using the rule that the subset contains an aromatic ring center 5.9 to 7.1 Å from a triple bond center. Approximately half of the compounds in this set also contain a positive charge center as shown by the shaded node labeled N101 in Figure 5.19. The structure of the pharmacophore is shown in Figure 5.20. This pharmacophore is consistent with experimental data that shows that propargylamines are monoamine oxidase inhibitors that operate by

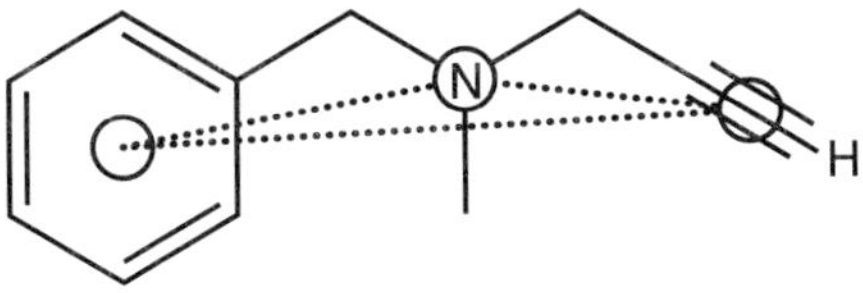

AL19120 (in node N101)

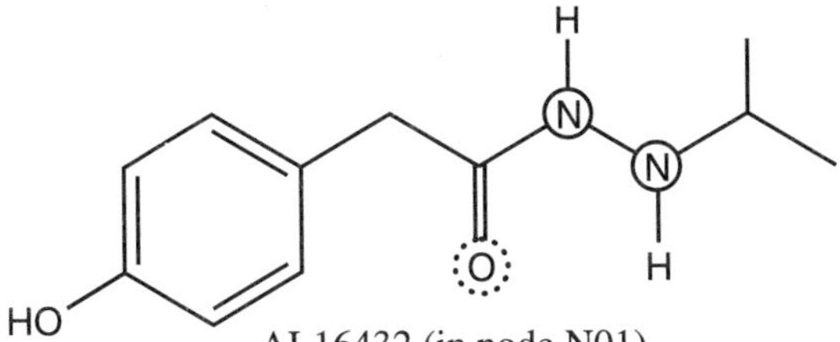

AL16432 (in node N01)

FIGURE 5.20 Pharmacophores derived from the tree. Chemical structures of the pharmacophores derived from the tree in Figure 5.19. The node designations correspond to the active nodes on the tree (N01 and N101). Reprinted with permission from Reference 6.

covalent bonding to a flavin cofactor.[6] A different pharmacophore is identified in node N01. This pharmacophore contains two hydrogen-bond donors adjacent to a carbonyl. The proposed mechanism for these inhibitors is hydrolysis to hydrazines that in turn are nonselective irreversible inhibitors. An important aspect of this case study is that the SAR rules were derived from a set of compounds that included several different chemical series. Pattern recognition involves identification of chemical features within and across the different series associated with activity.

Recursive partitioning can be performed on a multivariate vector of responses to correlate structural features with receptor or protein subtype selectivity. Figure 5.21 shows a simple recursive-partitioning tree based on using topological chemical descriptors[27] to discriminate between selectivity for three different mutants of an HIV protease. Each node on the tree contains a small picture of the selectivity profile of the compounds represented by the node.[28] Differences between the bar height in terminal nodes of the tree indicate that the structural feature associated with splitting the pair of nodes has an effect on producing a different selectivity profile.

Software packages that perform recursive partitioning, include: CART, C4.5, FIRM, KnowledgeSEEKER, and CHAID.[29-33] Recursive partitioning is an intuitive and robust method of tree regression. The particular versions of recursive partitioning found in FIRM and KnowledgeSEEKER use Bonferroni-adjusted hypothesis testing for combining observations and thus offer some protection from oversplitting the data.

Tree regression and specifically, recursive partitioning is a robust analysis method that is ideally suited for the analysis of large multivariate data. The analysis is easy to understand yet can analyze complex datasets that would normally require the use of neural networks or other nonlinear regression techniques.

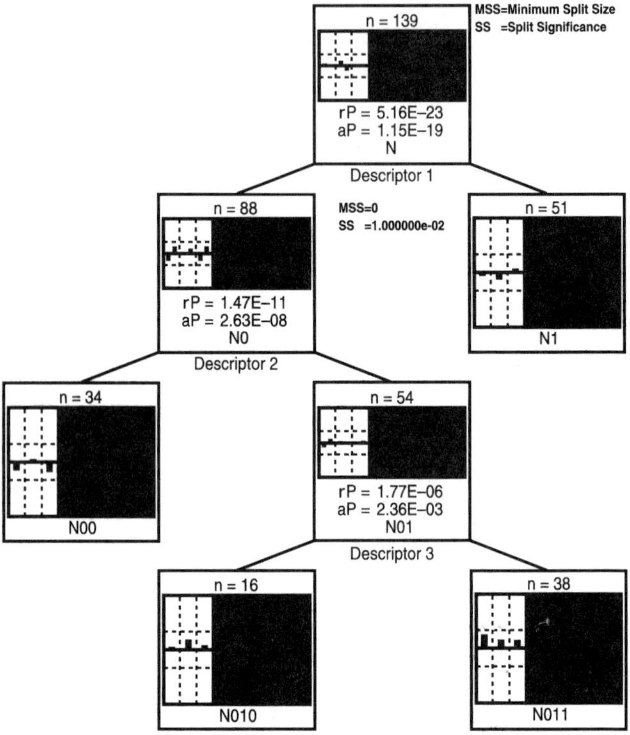

Subtype	Avg. pki.	SD (pki)	Min.	Max.
WT	8.43	0.92	6.27	10.77
M1	6.62	0.86	4.50	9.40
M2	6.41	0.73	4.70	7.82

FIGURE 5.21 Using multivariate responses with trees. Analysis of a multivariate response using recursive partitioning. This example shows a recursive partitioning tree for the affinity of a set of 139 different compounds for three mutants of an HIV protease. One structural feature is responsible for the first split of the tree. The 51 compounds in this node (N1) have slightly less affinity for the second mutant as evidenced by the bar in the second column. The left side of the tree is split further by other chemical features. The node at the far left (N00) has relatively more selectivity for the second mutant than the first and third mutants. The node at the far right (N011) contains compounds that do not show selectivity, the bars are of roughly equal height. Reprinted with permission from Reference 34.

5.10 APPENDIX

5.10.1 Index of Databases and Software Tools for Bioinformatics

5.10.1.1 Databases

WWW Virtual Library (Oak Ridge National Laboratory)
http://www.ornl.gov/TechResources/Human_Genome/genetics.html

NCBI: nucleotide, protein, 3D structures, expressed sequence tags, and genome database (GenBank), ENTREZ, BLAST, OMIM
http://www.ncbi.nlm.nih.gov/
EMBL
http://www.embl-heidelberg.de/
Genome database (GDB)
http://gdbwww.gdb.org/
G protein database
http://www.gpcr.org/7tm/
Mouse Genome Informatics
http://www.informatics.jax.org/
Protein sequences (SWISS-PROT)
http://expasy.hcuge.ch/sprot/sprot-top.html
Protein structures (PDB)
http://www.pdb.bnl.gov/
Nucleotide sequences (NAD)
http://ndbserver.rutgers.edu/

5.10.1.2 Software Tools

Sequence searching (BLAST)
http://ww3.ncbi.nlm.nih.gov/BLAST
NCSA Biology Workbench
http://biology.ncsa.uiuc.edu/
Gene structure prediction programs (exons, introns, promoters, poly-A sites)
http://CCR-081.mit.edu/GENSCAN.html
http://www.cs.jhu.edu/labs/compbio/veil.html
http://www.cs.jhu.edu/labs/compbio/morgan.html
http://www-hgc.lbl.gov/inf/genie.html
http://compbio.ornl.gov/Grail-bin/EmptyGrailForm
http://genomic.sanger.ac.uk/gf/gf.shtml
http://bmerc-www.bu.edu/geneid.html
Mathematical modeling and simulation software (DBSOLVE)
http://websites.ntl.com/~igor.goryanin

5.10.2 Other Resources

The Institute for Genomic Research
http://www.tigr.org/
The Sanger Centre
http://www.sanger.ac.uk/
Whitehead Institute/MIT Center for Genome Research
http://www.genome.wi.mit.edu/
Stanford human genome center
http://shgc.stanford.edu/

5.10.3 Human Genome Project Information

National Human Genome Research Institute at the National Institutes of Health
http://www.nhgri.nih.gov/
http://www.ornl.gov/

NCBI map of the human genome
http://www.ncbi.nlm.nih.gov/SCIENCE96

REFERENCES

1. Macauley, J., Wang, H., and Goodman, N., A model system for studying the integration of molecular biology databases, *Bioinformatics,* 14, 575–82, 1998.
2. Sandberg, M., Eriksson, L., Jonsson, J., Sjostrom, M., and Wold, S., New chemical descriptors relevant for the design of biologically active peptides. A multivariate characterization of 87 amino acids, *J. Med. Chem.,* 41, 2481–91, 1998.
3. Collantes, E.R. and Dunn, W.J., Amino acid side chain descriptors for quantitative structure-activity relationship studies of peptide analogues, *J. Med. Chem.,* 38, 2705–13, 1995.
4. Nakai, K., Kidera, A., and Kanehisa, M., Cluster analysis of amino acid indices for prediction of protein structure and function, *Protein Engineering,* 2, 93–100, 1988.
5. Hall, L.H., Mohney, B., and Kier, L.B., Structure information at the atomic level for molecular graphs, *J. Chem. Inf. Comput. Sci.,* 31, 76–82, 1991.
6. Chen, X., Rusinko, A., Tropsha, A., and Young, S.S., Automated pharmacophore identification for large chemical data sets, *J. Chem. Inf. Comput. Sci.,* 39, 887–96, 1999.
7. Lyman, W.J., Reehl, W.F., and Rosenblatt, D.H., *Handbook of Chemical Property Estimation.* American Chemical Society, Washington, DC, 1991.
8. Schena, M., Shalon, D., Davis, R.W., and Brown, P.O., Quantitative monitoring of gene expression patterns with a complementary DNA microarray, *Science,* 270, 467–70, 1995.
9. Schena, M., Genome analysis with gene expression microarrays, *BioEssays,* 18, 427–31, 1996.
10. Schena, M. et al., Parallel human genome analysis: microarray-based expression monitoring of 1000 genes, *Proc. Natl. Acad. Sci.,* 93, 10614–19, 1996.
11. DeRisi, J. et al., Use of a cDNA microarray to analyze gene expression patterns in human cancer, *Nature Genetics,* 14, 457–60, 1996.
12. Lockhart, D. et al., Expression monitoring by hybridization to high-density oligonucleotide arrays, *Nature Biotechnol.,* 14, 1675–80, 1996.
13. Velculescu, V.E., Zhang, L., Vogelstein, B., and Kinzler, K.W., Serial analysis of gene expression, *Science,* 270, 484–7, 1995.
14. Laing, P. and Pardee, A.B., Differential display of eukaryotic messenger RNA by means of the polymerase chain reaction, *Science,* 257, 274–80, 1992.
15. McAdams, H.H. and Shapiro, L., Circuit simulation of genetic networks, *Science,* 269, 650–6, 1995.
16. DeRisi, J.L., Bishwanath, R.I., and Brown, P.O., Exploring the metabolic and genetic control of gene expression on a genomic level, *Science,* 278, 680–6, 1997.
17. Novak, B. and Tyson, J.J., Numerical analysis of a comprehensive model of M-phase control in Xenopus oocyte extracts and intact embryos, *J. Cell. Science,* 106, 1153–68, 1993.

18. Goryanin, I., Hodgman, T.C., and Selkov, E., (1999) Mathematical simulation and analysis of cellular metabolism and regulation, *Bioinformatics,* 15, 749–58, 1999.
19. Solomon, M.J. et al., Cyclin activation of p34cdc2, *Cell,* 63, 1013–24, 1990.
20. Weinstein, J.N. et al., An information-intensive approach to the molecular pharmacology of cancer, *Science,* 275, 343–9, 1997.
21. Baldi, P. and Brunak, S., *Bioinformatics: The Machine Learning Approach,* MIT Press, Cambridge, MA, 1998.
22. Arrigo, P., Fariselli, P., and Casadio, R., Can functional regions of proteins be predicted from their coding sequences? The case study of G protein-coupled receptors, *Gene,* 221, 61–110, 1998.
23. Anzali, S., Gasteiger, J., Holzgrabe, U., Polanski, J., Sadowski, J., Teckentrup, A., and Wagener, M., The use of self-organizing neural networks in drug design, *Perspectives in Drug Discovery and Design,* 9/10/11, 273–99, 1998.
24. Liu, J., Neuwald, A., and Lawrence, C., Bayesian models for multiple sequence alignment and Gibbs sampling strategies, *J. Am. Stat. Assoc.,* 432, 1156–70, 1995.
25. Lawrence, C., Altschl, S., Boguski, M., Liu, J., Newald, A., and Wooton, J., Detecting subtle sequence signals: a Gibbs sampling strategy, *Science,* 262, 208–14, 1993.
26. Jonkers, R., Da Silva, A., Strosberg, A.D., Bouvier, M., and Marullo, S., New molecular and structural determinants involved in beta2-adrenergic receptor desensitization and sequestration, *J. Biol. Chem.,* 271, 9355–61, 1996.
27. Hawkins, D.M., Young, S.S., and Rusinko, A., Analysis of a large structure-activity data set using recursive partitioning, *Quant. Struct.-Act. Relat.,* 16, 1–7, 1997.
28. Keefer, C., Glaxo Wellcome, personal communication, 1998.
29. Hawkins, D.M. and Kass, G.V., Automatic Interaction Detection, in *Topics in Applied Multivariate Analysis,* D.M. Hawkins, Ed., Cambridge Univ. Press, Cambridge, MA, 1982.
30. Breiman, L., Friedman, J.H., Olshen, R.A., and Stone, C.J., *Classification and Regression Trees,* Wadsworth, California, 1984.
31. ANGOSS KnowledgeSEEKER for Windows Users Guide Version 3.0, ANGOSS Software International Limited, Toronto, Canada, 1994.
32. Kass, G.V., An exploratory technique for investigating large quantities of categorical data, *Appl. Statist.,* 29, 127–99, 1990.
33. Quinlan, J.R., *C4.5: Programs for Machine Learning,* Morgan Kaufman, San Mateo, California, 1993.
34. Lutz, M.W. and Kenakin, T.P., *Quantitative Molecular Pharmacology and Informatics in Drug Discovery,* John Wiley and Sons, Chichester, UK, 1999.

ADDITIONAL READING

Bishop, M.J., *Guide to Human Genome Computing,* Academic Press, Cambridge, UK, 1998.

Bywater, R., Vriend, G., Oliveira, L., and v. Aalten, D., Using sequence information and model building to explore subtype specificity in GPCRs, in *Protein Folds,* H. Bohr and S. Brunak, Eds., CRC Press, Boca Raton, FL, 1995.

Frishman, D., Heumann, K., Lesk, A., and Mewes, H., Comprehensive, comprehensible, distributed and intelligent databases: current status, *Bioinformatics,* 14, 551–60, 1998.

Kel, A., Ptitsyn, A., Babenko, V., Meier-Ewert, S., and Lehrach, H., A genetic algorithm for designing family-specific oligonucleotide sets used for hybridization: the G protein-coupled receptor protein superfamily, *Bioinformatics,* 14, 259–70, 1998.

Krogh, A., An introduction to hidden Markov models for biological sequences, in *Computational Methods in Molecular Biology,* S.L. Salzberg, D.B. Searles, and S. Kasif, Eds., Elsevier, Netherlands, 1998.

Kuipers, W., Oliveira, L., Paiva, A.C.M., Rippmann, F., Sander, C., Vriend, G., Kruese, C.G., van Wijngaarden, I., and Ijzerman, A.P., Sequence-function correlation in G protein-coupled receptors, in *Membrane Protein Models,* J. Findlay, Ed., BIOS Science Publishing, UK, 1995.

Kuipers, W., Oliveira, L., Vriend, G., and Ijzerman, A.P., (1997) Identification of class-determining residues in G protein-coupled receptors by sequence analysis, *Receptors and Channels,* 5, 159–74, 1997.

McGregor, M.J. and Muskal, S.M., Pharmacophore Fingerprinting. 1. Application to QSAR and focused library design, *J. Chem. Inf. Comput. Sci.,* 39, 569–74, 1999.

McGregor, M.J. and Muskal, S.M., Pharmacophore Fingerprinting. 2. Application to primary library design, *J. Chem. Inf. Comput. Sci.,* 1999.

Noble, D., Levin, J., and Scott, W., Biological simulations in drug discovery, *Drug Discovery Today,* 4, 10–6, 1999.

Schulze-Kremer, S., *Molecular Bioinformatics: Algorithms and Applications,* Walter de Grutyer: New York, 1996.

Sze, S-H., Roytberg, M.A., Gelfand, M.S., Mironov, A.A., Astakhova, T.V., and Pevzner, P.A., Algorithms and software for support of gene identification experiments, *Bioinformatics,* 14, 14–9, 1998.

6 Computer-Intensive Statistical Procedures

John Ludbrook

CONTENTS

6.1 INTRODUCTION

It might be thought that computer-intensive procedures were invented after the introduction of mainframe computers in the early 1950s or desktop computers in the early 1980s. For the most part, this is not so. Most of the statistical procedures to be discussed were invented in the 1930s and only one — bootstrapping — in the last 50 years. For instance, R.A. Fisher described his exact test for 2 × 2 tables in the first edition of his *The Design of Experiments,* published in 1935, in connection with the now-famous "thought" experiment about the lady who claimed she could tell whether milk or tea had been added first to a cup.[1] This test requires the tedious calculation of factorials, but tables of these for numbers up to 300 were provided in 1938 by Fisher and Yates.[2] Fisher also described in the first edition of *The Design of Experiments* an exact permutation test for the difference between paired means. This involved calculating no fewer than 32,768 possible permutations of the difference. Fisher did not get the outcome exactly right in the first edition, but in the second

0-8493-0100-9/01/$0.00+$.50

edition, published two years later, he did. An achievement at least as remarkable was made by Eden and Yates in 1933, when they analyzed a randomized block experiment by permutation.[3] The number of possible permutations was 24^8, but Eden and Yates overcame the computational challenge by an ingenious Monte Carlo pseudo-random sampling technique in which they used cards. Fisher also introduced the notion of maximum-likelihood estimation in the 1930s. Even bootstrapping, a concept described formally by Efron in 1979 that is perhaps the most computer-intensive procedure imaginable,[4] could be executed by hand, though less than optimally.

The foregoing supports the point that the near-universal availability of fast computers on desktops and laps is a convenience rather than a necessity. However, it must be admitted that they are very, very convenient. For instance, it took Fisher two years to correctly execute his permutation test on the same set of data it took me 0.33 seconds to execute on my desktop!

This leads to my issuing a word of warning. Almost every biomedical investigator has access to very powerful computers and can afford sophisticated statistical software. However, this has led to a reliance on the software manuals and "Help" files for selecting the appropriate statistical procedures to analyze one's data, and to faith that the software will execute these procedures correctly. Neither of these assumptions is justified. Biomedical investigators continue to choose inappropriate statistical procedures to analyze their data. There is ample evidence that software manuals (and even some well-known statistical texts) can be an unreliable source of advice when choosing the appropriate test, and the statistical software may sometimes execute tests incorrectly.[5]

The computer-intensive tests to be discussed reflect to some extent the preferences and prejudices of the author. The list of tests is also restricted to those that might be of interest to pharmacologists, whether those whose research is conducted in the laboratory or those whose field is clinical pharmacology. A final point is that biomedical investigators in general, and pharmacologists in particular, employ very small groups in their experiments.[6] This fact, plus the rarity with which random sampling is used compared with randomized designs in biomedical research, provide strong arguments in favour of using exact tests of significance rather than those based on assumptions about population distributions. Computer-intensive procedures, including those described as exact, compile an empirical sampling distribution that is specific to the particular dataset and against which the observed outcome of the experiment, or the observed test statistic, is matched.

6.2 A NOTE ON INFERENCE

6.2.1 THE CLASSICAL POPULATION MODEL

This has been the main basis for statistical theory over the past 70 years.[7,8] It depends on the assumption that the experimental groups have been acquired by random sampling of defined (though not necessarily accessible) populations. The sample sizes are very small compared with the size of the populations. Significance testing proceeds as follows. A population parameter of interest is selected, such as the mean, or the difference between means; the variance, or the variance ratio; or the difference between, or ratio of, proportions. The corresponding sample statistic is then calculated. A further assumption is then made about the distribution of the sample statistic in the population (for instance, according to the normal, t, F, or chi-squared distributions). A null hypothesis is then formulated. In its general form, this hypothesis states that the two or more random samples have been drawn from the same population. Somewhat more specifically, it may propose that the samples have been taken from populations with equal means, or equal variances, or equal proportions. The next step is to conduct an appropriate test of significance for the null hypothesis, such as Student's t test, Fisher's F test (analysis of variance), or one or another formulation of Pearson's χ^2 test. It is also necessary to set the desired level of significance. Traditionally, this is the 5% level ($P \leq 0.05$). What the 5% level of significance is intended to mean is that if a large (infinite) number of random samples were to be taken from the population(s) actually sampled in the experiment, with replacement, then in not more than one sample in 20 would the statistic of interest be equaled or exceeded. Hence the convention of expressing the level of significance as $P \leq 0.05$.

This lengthy and complex description of the classical, population, model of significance is given so as to be able to point out the difficulties with it in the setting of biomedical and pharmacological research. The fact is that investigators in these fields never, or almost never, fulfill even the first assumption. That is, they do not take random samples from defined populations.[6]

6.2.2 The Randomization Model of Inference

Fisher first introduced the notion of randomization in experimental design in 1936.[9] However, the randomization model of inference has not been nearly as well defined and debated by statisticians as the classical, population, model. It is hard to avoid concluding that this is because mathematical statisticians find that there is little opportunity to exercise their skills on this model. Nevertheless, there have been statisticians who have spoken out in favor of it, especially Otto Kempthorne and George Box.[10-13] There have also been biomedical experimenters who have written in favor of it.[6,14,15] Under this model, it is supposed that a nonrandom sample of experimental units (humans, animals, tissues, or cells) has been acquired. The members of this sample are then allocated at random to two or more conditions or treatment regimens. "Condition" could mean, for instance, rest or exercise, normovolaemia or hypovolaemia; and "treatment" could mean active treatment or placebo, a new treatment or an established treatment. These options are familiar to biomedical and pharmacological researchers. The experiment is then performed, and the results analyzed. The null hypothesis is no more than that the conditions or treatments have no differential effects on the groups with respect to a statistic such as the mean, proportion, or odds. There is no reference to a population and therefore no assumption is, or can be, made about the frequency-distribution of the population parameter corresponding to the statistic of interest. Instead, an empirical and unique sampling distribution of the statistic of interest is compiled by techniques such as permutation (see later). The P value is the frequency with which values of the statistic are equal to, or more extreme than, the experimentally determined value occurring in the empirical distribution. The statistical inference refers only to the nonrandom sample that was randomized, and not to any theoretical more general population. Biological inferences beyond those made from the experiment can be made only by non-statistical arguments about how representative the original nonrandom sample was.

6.3 ANALYSIS OF TABLES OF FREQUENCY BY EXACT METHODS

This section is concerned with computer-intensive or "exact" techniques for analyzing tables of frequencies. An excellent elementary account of some of these is given by Siegel and Castellan,[16] and a more advanced account by Agresti.[17] This seems a good place at which to start any essay on computer-intensive statistical methods because the concept of an exact test is relatively simple. The older and more conventional methods for analyzing tables of frequencies depend on tests based on the chi-squared or normal distributions. These are approximate or asymptotic tests of significance that are reasonably accurate for 2 × 2 tables in which the cell frequencies are large. They are notably inaccurate when the cell frequencies are small. They have been, or should be, replaced by tests that are based on the exact sampling distributions for specific tables.

6.3.1 Analysis of 2 × 2 Tables

Two-by-two tables (that is, 2 columns, 2 rows, sometimes called four-fold tables), are the simplest way of tabulating frequencies. A stereotype 2 × 2 table is shown in Table 6.1. The dataset that will be used as an example is in Table 6.2.[18]

The asymptotic Pearson χ^2 *test.* This is the oldest method for analyzing tables of frequencies. Karl Pearson described the chi-squared distribution in 1900,[19] though it was not until R.A. Fisher described the correct degrees of freedom in 1922 that the χ^2 statistic could be used as a practical test of significance.[20] It is usually described as testing for goodness of fit: that is, for identity of observed and expected frequencies. However, one should, whether fairly or not, regard Pearson's χ^2 test as of

TABLE 6.1
Stereotype 2 × 2 Table of Frequencies

	Column 1	Column 2	Total
Row 1	a	b	(a + b)
Row 2	c	d	(c + d)
Total	(a + c)	(b + d)	*N*

Note: The convention is to describe Rows as r_1, r_2, and so forth, and Columns as c_1, c_2, and so forth. It is usual to regard Columns as treatment groups, Rows as the outcomes of treatment.

TABLE 6.2
Example of a 2 × 2 Table after Davis[18]

	Group A	Group B	Total
Dead	3	2	5
Alive	6	19	25
Total	9	21	30

Note: For the purpose of discussion, assume that the investigators have acquired a nonrandom sample of 30 experimental units (humans, animals, tissues, cells). By a process of restricted randomization, they have assigned 9 to treatment Group A, 21 to treatment Group B. The outcomes of treatment are given by the Rows.

historical interest only. It is sufficient to cite Yates' recognition of its inaccuracy in his paper of 1934, when he prescribed his correction for continuity and controlled its efficacy by comparing outcomes with those of the Fisher exact test,[21] or to cite Cochran's paper of 1954 in which he suggested complex rules based on expected cell frequencies for defining the limits of cell frequencies for 2 × 2 and $r \times c$ tables below which the Pearson χ^2 test is unsafe.[22] Yet some biomedical investigators still analyze their results with this test. However, the Pearson χ^2 statistic can also be used in an exact test (see later).

The Fisher exact test. As indicated earlier, Fisher described this technique for analyzing 2 × 2 tables in 1935 in his *The Design of Experiments.*[1] It depends on constructing the hypergeometric distribution for all possible tables having the same marginal totals as the observed table. For any given table, and using the notation of Table 6.1, the probability of occurrence of the entries in the table is given by the formula:

$$\text{Pr} = \frac{(a+b)(c+d)(a+c)(b+d)}{N!a!b!c!d!}$$

where ! indicates factorial.

All possible probabilities of occurrence for the example in Table 6.2 are given in Table 6.3.

The null hypothesis for the Fisher test is rather vague. It is effectively that the cell frequencies in the observed table are disposed at random. The hypothesis is tested by summing the observed hypergeometric probabilities (Pr) with all more extreme (smaller) values of Pr (Table 6.3). If the summation is done unidirectionally, the test is one-sided. If a two-sided test is preferred (as it almost always should be), both tails of the distribution of Pr are summed to arrive at P values (Tables 6.3 and 6.4).

The Fisher test can be executed by means of a handheld calculator, but even for the example of Table 6.2 it is a tedious process. By means of a computer, the test can be performed on even the largest 2 × 2 table in a matter of milliseconds.

TABLE 6.3
List of All Possible 2 × 2 Tables Resulting from Example in Table 6.2, Given Fixed Marginal Totals

Tables				Hypergeometric	Test Statistics		
r_1c_1	r_2c_1	r_1c_2	r_2c_2	Pr	Exact χ^2	Exact G^2	Exact lnOR
5	4	0	21	0.00008842*	14.00*	14.67*	∝*
4	5	1	20	0.018568*	7.14*	6.63*	2.77*
3	**6**	**2**	**19**	**0.12378***	**2.57***	**2.37***	**1.56***
2	7	3	18	0.33599	0.29	0.27	0.54
1	8	4	17	0.37798	0.29	0.30	0.63
0	9	5	16	0.14279	2.57*	3.98*	∝*
	Total Pr for table:			**1.0000**			

Note: The example is from Davis[18]. The observed table is in bold. * indicates a value that is ≥ that observed in either direction. For the Fisher test, two-sided *P* is the sum of hypergeometric probabilities (Pr) that are equal to or less than that observed in both directions. For the exact χ^2, G^2, and lnOR tests, two-sided *P* is the sum of the hypergeometric probabilities corresponding to tables in which the test statistic is equal to or greater than that for the observed table, in both directions.

TABLE 6.4
Outcomes of Significance (Hypothesis) Testing on the Example of Table 6.2

Test	Exact *P* Values		Asymptotic *P* Values	
	1-Sided	2-Sided	1-Sided	2-Sided
Pearson χ^2	0.143	0.286	0.0544	0.109
Yates χ^2	NA	NA	0.143	0.285
Fisher exact	0.143	0.143	NA	NA
Odds ratio	0.143	0.286	0.0544	0.109
Likelihood ratio	0.143	0.286	0.0619	0.124
Relative risk (ratio of proportions)	NA	1.000	NA	0.115*
				0.109*
Difference of proportions	NA	0.260	NA	0.161
Permutation test on proportions (difference between proportions)	0.143	0.286	NA	NA

Note: P values from StatXact 3.1 (Cytel Software Corporation, Cambridge MA). * using two different formulae. NA, not applicable.

The odds ratio (OR). This is defined as:

$$\text{OR} = \frac{a / c}{b / d}$$

The null hypothesis is that OR = 1, a relatively specific one.

Its asymptotic form makes use of the fact that lnOR (the natural logarithm of OR) is distributed approximately as the normal distribution. However, as with the Pearson test, this approximation is poor when cell sizes are small. There is an exact form of the test that depends on calculating lnOR for all 2 × 2 tables with the same marginal totals as those observed. *P* for the null hypothesis is obtained by summing the probabilities from the hypergeometric distribution that correspond to values for lnOR equal to or greater than that observed (Tables 6.3, 6.4). One-sided values for *P* are obtained by summing in one direction, the preferred two-sided *P* by summing both tails.

The exact χ^2 test. The logical basis for this is similar to that of the odds ratio test. The χ^2 statistic is calculated for all 2 × 2 tables with the same marginal totals as those observed. *P* for the null hypothesis is obtained by summing the probabilities from the hypergeometric distribution that correspond to values for χ^2 equal to or greater than that observed (Table 6.3). One-sided values for *P* are obtained by summing in one direction, the preferred two-sided *P* by summing both tails.

Likelihood ratio. The log-likelihood statistic, sometimes called G or G^2, is described well by Sokal and Rohlf.[23] The G^2 statistic is distributed asymptotically and approximately according to the chi-squared distribution. It is usually used for log-linear modeling, and for testing the independence of three-way or multiway tables of frequency.[23] However, it can be used to evaluate single 2 × 2 tables under a null hypothesis involving goodness of fit that is as vague as that for the Pearson test. There is an exact form of the log-likelihood ratio test. The G^2 statistic is calculated for all 2 × 2 tables with the same marginal totals as those observed. *P* for the null hypothesis is obtained by summing the probabilities from the hypergeometric distribution that correspond to values for G^2 equal to or greater than that observed (Tables 6.3, 6.4). One-sided values for *P* are obtained by summing in one direction, the preferred two-sided *P* by summing both tails.

Ratio of, or differences between, proportions. At first sight it is appealing to use the ratio of proportions (often called relative risk or RR), or the arithmetic difference between proportions ($\rho_1 - \rho_2$), to test null hypotheses such as RR = 1, or $\rho_1 = \rho_2$. These are concepts that biomedical investigators feel comfortable with. There are, however, considerable difficulties in doing this, whether by asymptotic or exact techniques. There are algorithms for doing this asymptotically,[17] but they are not regarded as altogether satisfactory. There are also algorithms for conducting these tests exactly,[17] but these are not merely computer-intensive, but also difficult to interpret. In short, there is not much enthusiasm for using proportions.

There is, however, another way of looking at the difference between proportions that seems to get round the difficulties referred to above. First, set out the experimental data as a rectangular file, identifying groups (1, 2) and the outcomes as, for instance, 1 = dead, 2 = alive. Then the proportion for each group coincides with the mean value. Then a permutation test for equality of group means (see below) is identical to a test for equality of proportions. The outcome of such a test is given in Table 6.4.

Summary. The following points can be made about the analysis of 2 × 2 tables:

1. Nowadays, it is almost impossible to make a case for using anything other than exact, computer-intensive techniques for analysing 2 × 2 tables. Asymptotic methods that involve approximations to the chi-squared or normal distributions are almost invariably too liberal: that is, the resultant *P* value is too small and the Type 1 error rate inflated beyond that specified. The only possible exception is to use Yates' continuity correction to the Pearson asymptotic χ^2 test. This usually gives a somewhat more conservative outcome than the Fisher exact test, which it was designed to simulate.[21]
2. Which of the exact techniques is to be preferred? The exact form of Pearson's χ^2 test, the Fisher exact test, and the exact log-likelihood ratio have the weakness that the null hypothesis is rather vague. The tests have been described in terms of "goodness of fit" or "independence of rows and columns." They do not test for equality of proportions, though this is a not uncommon misapprehension. There are exact tests for equality of proportions: the relative risk (RR, or ratio of proportions) and the difference between proportions, the null hypotheses being quite specific. That is, they are that RR = 1 and $\rho_1 - \rho_2 = 0$. However, the algorithms for executing these tests exactly are unconditional: that is, it is assumed that neither the column nor the row marginal totals are fixed in advance. This can lead to some rather curious results (Table 6.4). The author believes that the best option is the test on odds ratios where the null hypothesis is OR = 1. The algorithm is a conditional one, the specific condition being that the column marginal totals are fixed in advance by the design of the experiment. When columns indicate groups and rows indicate outcomes, this conditioning is in accord with the usual experimental design, when group sizes (column totals) are determined in advance by the investigators. A neglected, but apparently valid way of testing for equality of proportions is to use a permutation test.
3. In the author's view, two-sided tests should be the norm in biomedical research in general, and pharmacological research in particular. However, it is worth noting that whereas the Fisher exact test, the Pearson exact test, and the exact tests on the odds ratio and log-likelihood ratio always result in identical one-sided values of *P*, the two-sided *P* value from the Fisher test is often less than that for the others (Tables 6.3, 6.4).[18]

6.3.2 2 × c Tables of Frequencies

The advantages of exact, computer-intensive tests over asymptotic tests are rather less as the number of rows and columns increases. Nevertheless, exact tests are always more accurate, especially if the expected frequencies in some cells are small.[17] The example used in this section was described by Cochran.[22]

Pearson's χ^2 and log-likelihood tests for independence of rows and columns. These can be executed in the familiar, asymptotic form or in an exact form. Outcomes from the example are given in Table 6.5, and are very similar.

Fisher-Freeman-Halton exact test. This extends the Fisher exact test to 2 × c tables.[24] The outcome for the example is given in Table 6.5.

Cochran-Armitage test for ordered 2 × c tables. Almost simultaneously, Cochran and Armitage described an asymptotic method of partitioning the χ^2 statistic in the case of 2 × c tables when the columns represent a variable that increases (or reduces) in a linear fashion.[22,25] Genuine linearity is not necessary, and rarely occurs with categorical variables, but monotonicity is essential. The test can be executed in either an asymptotic or exact form. Its value is illustrated by the example (Table 6.5). If the columns are regarded as in no particular order so that the null hypothesis is merely one of independence of rows and columns, P from the above tests ranges between 0.139 and 0.144. However, if the columns represent a monotonically ascend-

TABLE 6.5
Example of a 2 × c Table of Frequencies from Cochran[22]

Skin Infiltration	Col 1	Col 2	Col 3	Col 4	Col 5	Total
Extensive	1	13	16	15	7	52
Slight	11	53	42	27	11	144
Total	12	66	58	42	18	196

Note: The table summarizes observations on lepers undergoing treatment with sulphones and streptomycin. The rows indicate the severity of skin infiltration before treatment. The columns indicate the clinical responses to treatment.

If the clinical responses are in no particular order, the outcomes of appropriate tests in terms of P values are:

Asymptotic Pearson χ^2 test	0.142
Exact Pearson χ^2 test	0.142
Asymptotic log-likelihood ratio test	0.123
Exact log-likelihood ratio test	0.139
Fisher-Freeman-Halton exact test	0.144

However, in reality the columns are in monotonic ascending order in terms of clinical response: 1 = Worse. 2 = Unchanged. 3= Slight improvement. 4 = Moderate improvement. 5 = Marked improvement. If the goal is to test for an association between the initial skin condition and the effect of treatment, then the Cochran-Armitage test is the appropriate one. The outcome is:

Asymptotic Cochran-Armitage test	0.010
Exact Cochran-Armitage test	0.011

ing order (as was the case in Cochran's example), P from the asymptotic or exact forms of the Cochran-Armitage test are 0.010 and 0.011, respectively. The null hypothesis is, of course, different: expressed in a two-sided fashion it is that there is no monotonic change in the proportions according to columns.

Summary. The following points can be made about the analysis of 2 × c tables:

1. For unordered 2 × c tables, the Fisher-Freeman-Halton exact test is to be preferred.
2. For ordered 2 × c tables, the exact form of the Cochran-Armitage is the procedure of choice.

6.3.3 Unordered r × c Tables of Frequencies

When there are more than two rows in an r × c table, the null hypothesis is of necessity quite nonspecific. It is merely that there is no association (interaction) between rows and columns. In tables such as these, the advantage of exact over asymptotic tests is less obvious, unless there are several cells in the table with low expected frequencies. The tests available are as follows.

Pearson χ^2 test, asymptotic or exact versions. Log-likelihood ratio test, asymptotic or exact. The properties of these have already been described under 2 × c tables.

Fisher-Freeman-Halton exact test. This has the same properties as for 2 × c tables.

Summary. For unordered r × c tables, there is really little choice but to use the exact Fisher-Freeman-Halton test.

6.4 PERMUTATION (RANDOMIZATION) TESTS

As indicated earlier, the idea of permutation tests goes back to Fisher, and Eden and Yates in the 1930s. There are now excellent books on this topic, in particular Edgington[15] and Manly.[26] These two books deal with permutation tests in the context of the randomization model of inference, and are directed towards investigators in the biological and biomedical sciences. Other books deal with the topic of computer-intensive tests in a more pragmatic fashion, without being based on explicit models of inference.[27,28] There is also an article that deals with these tests in the setting of biomedical research.[6]

The logic that underlies these tests is remarkably simple, and falls within the framework of the randomization model of inference. That is, the null hypothesis is that there is no differential effect of treatments on the randomized groups. The goal is to compile the permutation (randomization) distribution for each set of experimental data. The only assumption is that there has been a randomized experimental design. All possible permutations of the experimental data are then compiled, preserving the actual group numbers and sizes. Then for each permutation the statistic of interest is calculated. The most commonly used statistic, and the easiest to understand, is the difference between two means. However, when there are more than two groups, the F statistic (or a simplified version of it) can be calculated as for a one-way or two-way analysis of variance. The values of the statistic associated

TABLE 6.6
The Process of Exact Permutation for Two Independent or Two Related Groups

Two Independent Groups

Formula for calculating all possible permutations for two independent groups of size n_1 and n_2:

$$\frac{(n_1 + n_2)!}{(n_1)!(n_2)!}$$

The 10 possible permutations for $n_1 = 2$, $n_2 = 3$. Observed values for n_1 were A, B, and for n_2 were C, D, E.

A	C	A	B	A	B	A	B	B	A
B	D	C	D	D	C	E	C	C	D
	E		E		E		D		E

B	A	B	A	C	A	C	A	D	A
D	C	E	C	D	B	E	B	E	B
	E		D		E		D		C

Two Related Groups (Matched Pairs)

Formula for calculating all possible permutations of two related groups of size n is 2^n.

The 8 possible permutations for group size n. Observed values for the pairs were A1 & A2, B1 & B2, C1 & C2.

A1	A2	A2	A1	A1	A2	A1	A2
B1	B2	B1	B2	B2	B1	B1	B2
C1	C2	C1	C2	C1	C2	C2	C1

A2	A1	A2	A1	A1	A2	A2	A1
B2	B1	B1	B2	B2	B1	B2	B1
C1	C2	C2	C1	C2	C1	C2	C1

with the permutations are then placed in rank order. The P value is calculated as all values of the selected statistic that are equal to or greater than that observed. This is done for both tails of the permutation distribution for a two-sided P, and for one tail to obtain a one-sided P.

As will become clear, the number of possible permutations increases steeply with the number of groups and the group sizes. Even with a very fast computer it may be quite impractical to list all possible permutations. In this case, a Monte Carlo pseudo-random sample of all the possible permutations is taken, usually 10,000. The calculation of P is then done on this Monte Carlo sample (and confidence limits for P can be provided).

There are several computer programs for executing permutation tests, listed in Section 6.6.

Two independent groups. How all possible permutations are compiled is illustrated in Table 6.6. The formula for calculating the number of possible permutations

TABLE 6.7
Number of Possible Permutations for Two Independent or Two Related Groups

Two Independent Groups		Two Related Groups (Matched Pairs)	
Size: $n_1 = n_2 =$	Permutations	Size: $n =$	Permutations
5	252	5	32
10	184,756	10	1,024
15	155,117,520	15	32,768

Note: Number of permutations calculated from the following formulae. Two independent groups: $(n_1 + n_2)!/(n_1)!(n_2)!$, where ! indicates factorial. Two related groups (matched pairs): 2^n.

is $(n_1 + n_2)!/(n_1)!(n_2)!$ where n_1 and n_2 are the group sizes, and ! indicates factorial. The rate of increase of permutations increases steeply with group size (Table 6.7), so that with groups larger than about 12, the Monte Carlo approach described above has to be used. The exact permutation distribution for the difference between two independent means is displayed in Figure 6.1. It should be noted that its bimodal appearance is very different from that of the t distribution.

Two related groups (matched pairs). How all possible permutations are compiled is shown in Table 6.6. The formula for the number of possible permutations is 2^n. The rate of increase with group size is much less than that for two independent groups (Table 6.7).

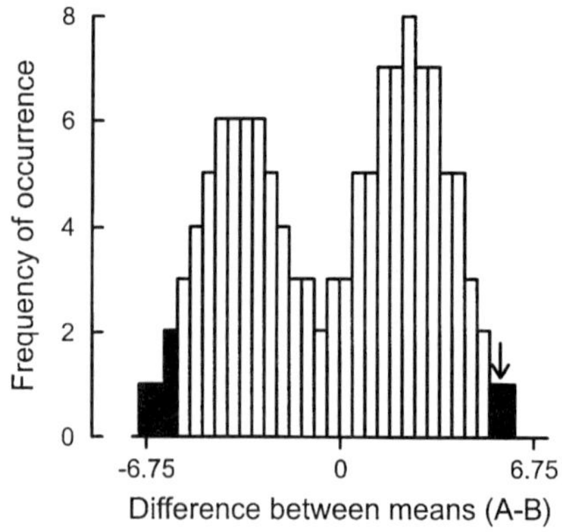

FIGURE 6.1 Frequency-histogram of the exact permutation distribution for the difference between two independent means. Values for group A: 20, 10, 9, 8, 6. Values for group B: 7, 5, 4, 3. There are 126 possible permutations. ■ the 6 permutations for which the difference between means equals or exceeds, in either direction, the observed value of 5.85, so that two-sided $P = 6/126 = 0.048$. ➡ observed value. From Ludbrook[41] with kind permission of Blackwell Science Asia.

Rank permutation tests. Wilcoxon described these in 1945,[29] and Mann and Whitney described exactly equivalent tests in 1947[30] producing the Wilcoxon signed rank-sum test for matched pairs and the Wilcoxon-Mann-Whitney test for two independent groups. Wilcoxon made it clear that he strongly supported Fisher's idea of permutation tests,[29] but in 1945 there were no computers with which to execute them, so they were altogether impractical. He recognized that if ranks were substituted for continuous variables, the computational problem was much simplified. The number of possible permutations remained the same, but the number of values assumed by the test statistic was very much smaller. However, it is the author's view that the rank permutation tests should no longer be used. The arguments in support of this view are given in Ludbrook and Dudley[6] and Bergmann et al.[5] Perhaps the strongest argument is that these are tests for differences in mean ranks, and under the ranking system used, this does not coincide with differences between medians (though this is a common misapprehension).

Permutation tests for multiple randomized groups. These correspond to one-way or two-way analysis of variance (ANOVA). The principle of the test is very similar to that for two groups, except that the statistic of interest is the F statistic (or, rather, a simplified version of it).

Eden and Yates were the first to describe a statistic equivalent to F,[3] and Edgington gives a simple explanation of this.[15] The equivalent test statistic is:

$$\sum (T_i^2 \,/\, n_i)$$

where T is the sum of all values in a particular group, n the group size. T^2/n is summed across all groups under consideration, for each permutation of group values. If all groups are of the same size, the expression can be reduced to:

$$\sum (T_i)^2$$

These statistics are equivalent to F in the sense that across all the permutations, the rank order of $\sum (T_i^2 \,/\, n_i)$ is identical to that of F. Obviously, the equivalent test statistic requires much less computation than F.

There are no problems with executing an exact or Monte Carlo version of one-way ANOVA. However, two-way ANOVA and factorial analysis by permutation present some problems. The main effects can be extracted without difficulty. But more often than not, investigators use interactions within two-way ANOVA to test their main hypothesis. Interactions do present a problem of execution by permutation. There is no consensus about the best way to go about extracting these,[15,26-28] and the outcomes using different techniques can be very different. Repeated measures ANOVA, commonly used in biomedical laboratory research, seem to have no corresponding permutation version, especially for within-group interactions.[31] This is scarcely surprising, since measurements made serially cannot be regarded as randomized. However, there may be no cause for despair. I know of no study that proves the case, but it is my strong impression from analyzing experimental data that the greater the number of randomized groups, regardless of their size, the less the advantage of permutation tests compared with classical F tests.

The use of permutation is not confined to comparing two or more group means. For instance, it can be used to test for equality of group variances. Its use in many other experimental designs has been described, such as for regression and correlation, time series, spatial analysis, repeated measures analysis, multivariate analyses, and clinical experiments of the $n = 1$ type.[15,26] However, the experiments used as examples for the above applications do not always fulfill the requirement of a randomized design. In the author's opinion, if this should be the case, then permutation techniques should not be used.

Summary. The following points can be made about the use of permutation (randomization) tests:

1. These should be used only when the experiment is a prospective one in which there has been genuine randomization of the experimental units (humans, animals, tissues, cells) to "treatment" groups.
2. If the circumstances are as above, and randomization has been to only two "treatments," a very strong and almost irrefutable case can be made for *never* using Student's t test, but only using a permutation test for equality of means (independent groups or matched pairs).
3. If the experimental design is one of randomization, then the best test for equality of variances is a permutation test.
4. The case for a permutation test is weaker if there are more than two randomized experimental groups, and the design corresponds to two-way or multiway analysis of variance or factorial analysis. There are two reasons for this view: (a) More often than not investigators are interested chiefly in interactions, such as a dose × treatment interaction when doses are given in random order and treatments are, for instance, exposure to one or more antagonists and a control. It is not at all clear what the best approach is for testing interactions by permutation. (b) There is little or no empirical evidence that, in complex designs such as these, permutation analysis provides greater accuracy in hypothesis-testing than conventional forms of analysis of variance. In other words, the greater the number of experimental groups and the greater the degrees of freedom attached to the Residual Mean Square in ANOVA, the closer the outcomes. This is especially so if the investigators (as they always should) incorporate an analysis of residuals into their analyses of variance and, if indicated, undertake data transformation or weighting.[32]
5. There seems to be no case at all for using permutation techniques to analyze data that have been acquired neither by randomization nor by random sampling. A very obvious example is when serial observations have been made in time, dose, or space when conventional repeated-measures analysis of variance would be used with appropriate correction for serial autocorrelation.[31]

6.5 MONTE CARLO METHODS

The term "Monte Carlo" originates from the gaming tables of Monaco. It refers to repeated random sampling of mathematically defined populations, with replacement. Thus, repeated tosses of a coin, repeated throws of a die, or repeated spins of a roulette wheel are forms of repeated random sampling with replacement, in which the sample size is 1. If one wishes to make inferences from Monte Carlo random sampling, they would clearly be under the population model of inference (see Section 6.2).

If Monte Carlo random sampling is applied to the results of biomedical experiments, it is likely that the sample size will be a good deal greater than 1. However, in other respects, the situation is not dissimilar to that of the games of chance referred to above:

1. The mathematical form of the parent population is known or postulated by the investigators.
2. The biological population is large compared with the sample size(s).
3. The experimental design involves genuine random sampling of the biological population.

Usually, hypothesis-testing or the construction of confidence intervals would be undertaken by parametric tests, for instance, those in which the t or F statistics are employed to analyze continuous data sampled from a population that is regarded as conforming to the normal distribution. So how can Monte Carlo methods be applied usefully in biomedical work?

A better question is how Monte Carlo methods have been applied. A search of Medline for the keyword "Monte Carlo" over the two-year period 1998–99 resulted in nearly 1000 references. Most of these referred to organ imaging via nuclear medicine techniques, computerized tomography (CT), magnetic resonance (MR) imaging, and positron emission tomography (PET). Many were concerned with radiotherapy. A smaller number were in the field of genetics. There were some concerned with clinical trials, but only a handful dealt with pharmacology, mostly pharmacokinetics. An exception was the paper by Christopoulos.[33]

6.5.1 Parametric Applications of Monte Carlo Methods

Monte Carlo simulation studies. One of the most common applications of Monte Carlo methods is in simulation studies.[34] These have been used extensively in the computer era to decide how accurate conventional tests such as t and F tests are when the randomly sampled population is not normal, or when the populations sampled have different variances, or both.[6] This process can be most easily explained if, for instance, the accuracy of 95% confidence intervals (95% CI) for a population mean are estimated by the conventional method of using the t statistic. It is illustrated in Table 6.8.

There is nothing magical about the figure 1000 for the number of Monte Carlo random samples (Table 6.8), though this is commonly used. Nor is there anything magical in repeating the process 1000 times (Table 6.8), though this, too, is a

TABLE 6.8
Flow Chart to Demonstrate the Process of Monte Carlo Simulation, to Test the Accuracy of Parametric *t*-Statistic-Based 95% Confidence Intervals (CI)

Take a mathematically definced population distribution

↓	↓
Take random sample of size *n*	Take 1000 Monte Carlo random samples of size *n*
↓	↓
Use *t* statistic to calculate 95% CI for population mean	Arrange means of 1000 samples in rank order and declare the 95% CI for population to be the 25th and 975th ranked values
	↓
	Repeat this step, say, 1000 times

↓

See whether the *t*-statistic-based CI lies within 95% confidence limits for the CI ascertained by Monte Carlo random sampling

Note: The process can be repeated with a variety of mathematically defined frequency distributions and with a range of sample sizes.

common choice. The accuracy of using the *t* statistic to estimate the 95% CI from a single sample can then be compared with the mean value from Monte Carlo sampling (and its 95% range). It should be apparent that the Monte Carlo method is precisely the theoretical process that underlies inferences made under the population model (see Section 6.2). The process illustrated in Table 6.8 can be repeated with different sample sizes (for instance, $n = 10$, 20, or 30). In the context of testing the "robustness" of 95% CIs for the population mean estimated by using the *t* statistic, populations with different mathematically defined frequency distributions can be sampled. These might be, for instance, distributions such as the normal, lognormal, Cauchy, beta, triangular, and so forth (see Reference 35 for definitions and illustrations). At the end of the simulation study, it should be possible to reach conclusions about the accuracy of the *t* statistic technique under different forms of nonnormality and different size samples. If the population distribution is genuinely normal, the *t* statistic approach is not likely to be subject to systematic error. Conversely, other population distributions and other patterns of sample size or variance may be associated with systematic errors in the estimation of the 95% CI.

The example given above is a rather trivial one, but Monte Carlo simulation studies with which the author is familiar have been used to test the accuracy and power of hypothesis-testing by classical *t* tests and analysis of variance (ANOVA),[6] the relative accuracy of different corrections for autocorrelation in repeated-measures ANOVA,[31] and the accuracy of control of the familywise Type I error rate by a wide variety of multiple comparison procedures.[36] From these simulations, it has been possible to make certain generalizations. For instance, the independent-sample *t* test is relatively robust against nonnormality of the population, but is very sensitive to inequality of population variance, in the direction that if the smaller sample comes from the population with the greater variance, the Type I error rate is inflated; whereas if the converse is the case, the statistical inferences are too conservative and the Type II error rate is inflated. Monte Carlo simulations have also been used to test the accuracy of bootstrapping (see below).

I have glossed over just how Monte Carlo random sampling is done, but it is described in relatively simple terms by Morgan.[34] For those in the know, it is a relatively simple matter to write computer programs (or to use programs written by others) to undertake the process described above.

Hypothesis-testing by Monte Carlo methods. It should be clear from the flow-chart given in Table 6.8, that all parametric tests of significance could be replaced by Monte Carlo tests. That is, if investigators must postulate that their biological populations conform to the normal distribution and would usually use the t or F statistic to test hypotheses, then multiple (Monte Carlo) random sampling of the theoretical normal distribution could be used to test the hypotheses, and this approach would be entirely in line with the theory of the population model of inference. However, this would be like using a piledriver to crack a nut. Not only would the process be highly computer-intensive, but ultimately its validity depends on knowledge of the frequency distribution of the biological population. This is impossible to discover, except by studying multiple random samples from that population. Furthermore, it requires that the biological population has been randomly sampled, a condition that is excessively rare in practice.[6] Nevertheless, the Monte Carlo approach has been supported both by biostatisticians and by pharmacologists. For instance, Manly gives examples of applying the method to the spatial distribution of plants and to mandible length in jackals.[26] Concato and Feinstein indicate applications in clinical research.[37] More immediately relevant, Christopoulos has applied the method to the analysis of ligand-receptor interactions.[33]

Nonparametric applications of Monte Carlo methods. One of the few revolutionary ideas in statistics in the computer era (post-1950) was the notion of bootstrapping introduced by Efron in 1979.[4] The word "bootstrapping" originates from one of the extravagent tales of Baron Munchausen, in which the hero extricates himself from a perilous situation by pulling himself up by his bootstraps. Bootstrapping has more or less eclipsed an earlier application of Monte Carlo methods known as "jacknifing," because of the greater generalizability of the former.

A Medline search for 1998–99 turned up only 46 articles in which bootstrapping was used. Only two of these were in the field of pharmacology. An earlier paper by Lew and Angus used bootstrapping in the context of nonlinear regression.[38]

The concept that underlies bootstrapping is relatively simple. It will be recalled that, under the population model of inference (Section 6.2), repeated, relatively small, random samples taken with replacement from a defined population allow the distribution of that population to be accurately constructed. However, biomedical populations are rarely, if ever accessible and so cannot be repeatedly randomly sampled. Efron proposed that if the actual, experimental, random sample were repeatedly randomly sampled with replacement, then an reasonably close approximation of the population distribution could be deduced.[4] Authoritative accounts of

TABLE 6.9
The Process of Bootstrap Resampling for Two Independent Samples

For the actual experimental samples, the values were A, B, C, D, E and F, G, H, I, respectively. The values in the first 10 bootstrap random resamples were as below.

C	H	B	G	D	H	E	G	E	F
C	G	D	G	B	F	D	G	C	F
B	H	D	G	B	F	B	H	B	H
E	H	E	F	A	F	C	I	D	I
D	A	A	D	E					
E	G	A	G	E	H	A	G	C	I
A	I	B	I	A	G	A	H	C	I
B	I	C	F	E	G	A	F	E	F
C	H	E	F	C	F	D	H	A	I
E	D	C	A	A					

Note: Because bootstrap resampling is done *with replacement*, the same values can appear more than once in each group. This is in contrast to compiling permutations (Table 6.6), where random resampling is done *without replacement*.

bootstrapping are given by Efron,[4] Efron and Tibshirani,[39] Mooney and Duval,[40] Manly,[26] and an elementary account by Ludbrook.[41]

The main thrust of Efron's proposal was to be able to construct bootstrap standard errors and confidence intervals without having to make assumptions about the population distribution. The simplest way of doing this is to compile, say, 1000 resamples of the original sample by random sampling with replacement. These are then placed in rank order in terms of the statistic of interest, for instance the mean, or the difference between two means. This is the so-called bootstrap percentile method. In Table 6.9 an example is given of the first 10 out of 1000 resamples of a set of data, and Figure 6.2 shows the bootstrap frequency distribution of the difference between two means for the same set of data as for Figure 6.1. It should be noted that there is no more than a hint of the bimodality that is evident in the permutation distribution of Figure 6.1. However, it is now clear on theoretical grounds and as a result of Monte Carlo simulation studies that there can be serious errors in confidence intervals constructed in this way from small samples (see Reference 41). There may be bias, in that the bootstrap statistic does not correspond to the population parameter (for instance, mean or difference between means). More importantly, the confidence interval is likely to be too narrow. Several ways of overcoming these deficiencies have been proposed. These include the percentile-t method,[39,42] and the bias-corrected and accelerated (BC_a) method.[39] The BC_a method is more widely applicable, but the percentile-t method is much simpler to execute (see Reference 41). Several other much more complex techniques for improving the

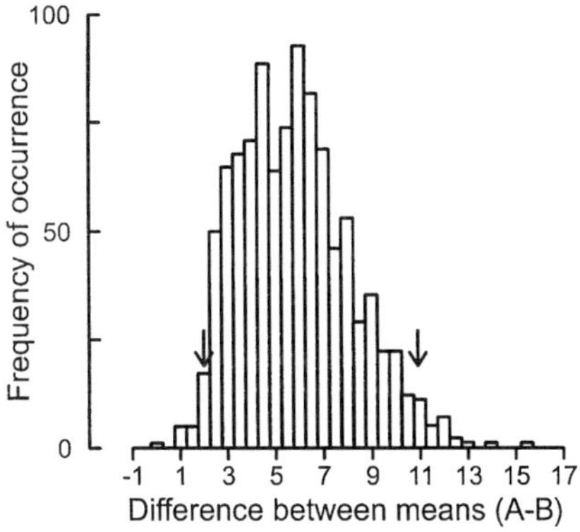

FIGURE 6.2 Frequency-histogram of the percentile bootstrap distribution for the difference between two independent means, based on 1000 resamples. Observed values as in Figure 6.1. Observed difference between means 5.85. ➧ indicate 95% bootstrap percentile confidence limits. From Ludbrook[41] with kind permission of Blackwell Science Asia.

accuracy of bootstrap resampling have been proposed. These include balanced random sampling,[39] double-bootstrapping,[40] and applicable only to very small samples, complete enumeration of all possible bootstrap resamples.[44] These proposals have yet to be fully evaluated.

Bootstrapping can also be used for hypothesis testing. In its crudest form, for testing for equality of population means, the bootstrap distribution is used in much the same sort of way as the permutation distribution (see Section 6.4). That is, the bootstrap differences between means (say 1000) are placed in rank order, and P is calculated as:

$$\frac{\text{Number of bootstrap mean differences} \geq \text{that observed}}{\text{Total number of bootstrap differences}}$$

However, this simple approach is susceptible to serious errors, especially when the original samples are small. Several solutions to this have been proposed, including the use of t- or F-like statistics,[41] but none is entirely convincing. In particular, a simulation study suggests that if samples are small (15 to 25) the percentile-t method is associated with an inflated risk of Type 1 error.[43]

The percentile-t bootstrap distribution for the dataset of Figures 6.1 and 6.2 is shown in Figure 6.3. It should be noted that it resembles the permutation distribution of Figure 6.1 in being bimodal.

Summary. The folllowing points can be made about Monte Carlo methods:

1. Strictly, these are applicable only to experimental situations in which genuine random samples are taken from defined populations. This is uncommon in biomedical research.
2. Monte Carlo simulation studies have played an important part in defining the conditions under which parametric or nonparametric methods for

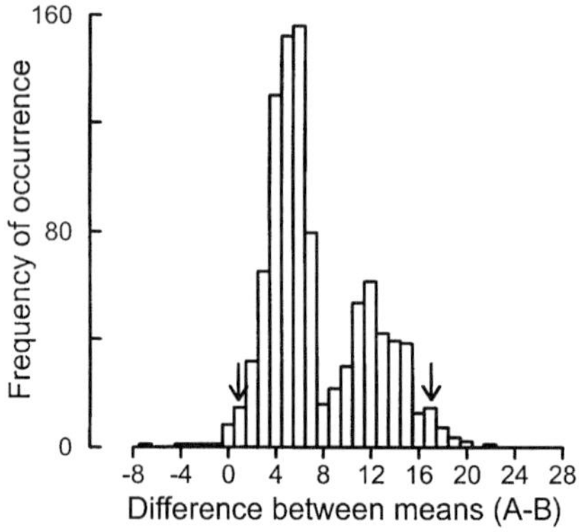

FIGURE 6.3 Frequency-histogram of the percentile-t bootstrap distribution for the difference between two independent means, based on 1000 resamples. Observed values as in Figure 6.1. Observed difference between means 5.85. ➡ indicate 95% bootstrap percentile-t confidence limits. Note the difference between this frequency-histogram of the percentile-t bootstrap distribution and that of the percentile distribution in Figure 6.2. From Ludbrook[41] with kind permission of Blackwell Science Asia.

hypothesis testing are accurate (or, more importantly, when they are potentially inaccurate).

3. Parametric Monte Carlo methods for testing hypotheses suffer much the same defects as more traditional methods. That is, both depend on assumptions about random sampling, and about the frequency distributions of the populations that have been randomly sampled.
4. Nonparametric Monte Carlo methods in general, and bootstrapping in particular, can be grossly inaccurate in estimating the dispersion of populations and in testing hypotheses when the experimental sample sizes are small. "Small" appears to refer to samples of less than 30.

6.6 PROGRAMS FOR COMPUTER-INTENSIVE TESTS

This should not be regarded as a comprehensive list, since it includes only programs of which the author is aware.

6.6.1 Categorical Data

A very comprehensive set of routines is found in StatXact (Cytel Software Corporation, Cambridge, MA). Some of these same routines are included as special modules in SPSS (SPSS Inc., Chicago) and SAS (SAS Institute Inc., Cary, NC).

6.6.2 Permutation Tests

DOS routines are provided by Edgington in his book.[15] Manly has a set of routines in the low-cost DOS program RT, described in his book,[26] and available from the Centre for Applications of Statistics and Mathematics, University of Otago, P.O.

TABLE 6.10
Outcomes of Statistical Analysis of the Dataset of Figures 6.1–6.3

Technique	95% Confidence Limits	95% Confidence Interval	Two-Sided *P* Value
Pooled-variance *t* test (raw values)	–0.93 to 12.63	13.56	0.081
Separate variance *t* test[a] (raw values)	NA	NA	0.073
Pooled variance *t* test (log transformed values)	NA	NA	0.028
Bootstrap percentile	2.25 to 11.0	8.75	0.014
Bootstrap percentile-*t*	1.18 to 17.14	15.96	0.022
Exact permutation	0.95 to 15.13	14.18	0.048

[a] Welch-Satterthwaite procedure. NA, not applicable. Dataset was group A: 20, 10, 9, 8, 6, group B: 7, 5, 4, 3. Observed difference between means: 5.85. Note that the "raw" bootstrap percentile method gives a CI that is far too narrow and a *P* value that is too small. Note also that the *t* test on log values does not test for $\bar{x}_1 - \bar{x}_2 = 0$ but $\bar{x}_1/\bar{x}_2 = 1$.

Box 56, Dunedin, New Zealand. The commercial program StatXact (Cytel Software Corporation, Cambridge, MA) has powerful routines for permutation tests on continuous and ranked data on a Windows platform.

6.6.3 Monte Carlo Methods

Routines for Monte Carlo simulations are usually written by the authors of the papers in which the results are described.

Routines for bootstrapping are described by the authors of several monographs.[39,40,45] They can also be constructed by those who are adept at using statistics programming languages such as S-PLUS (MathSoft Inc., Seattle, see Reference 44) and SAS (SAS Institute Inc., Cary, NC), and those who are adept at writing macros in programs such as Minitab (Minitab, State College, PA). There are some built-in routines in the most recent version of SYSTAT (SPSS Inc., Chicago). However, the occasional bootstrapper (like the author) is advised to use a spreadsheet. These carry out random resampling with replacement with impressive speed.

REFERENCES

1. Fisher, R.A., The design of experiments, in *Statistical Methods, Experimental Design, and Scientific Inference,* J.H. Bennett, Ed., Oxford University Press, Oxford, 1990.
2. Fisher R.A. and Yates, F., *Statistical Tables for Biological, Agricultural and Medical Research*, 6th edition, Harlow, Longman, London, 1963.
3. Eden, T. and Yates, F., On the validity of Fisher's *z* test when applied to an actual example of non-normal data, *J. of Agric. Sci.,* 23:6–16, 1993.
4. Efron, B., Bootstrap methods: another look at the jacknife, *Ann. of Stat.,* 7:1–26, 1979.

5. Bergmann, R., Ludbrook, J., and Spooren, W.P.J.M., Different outcomes of the Wilcoxon-Mann-Whitney test from different statistics packages, *The American Statistician,* 2000 (in press).
6. Ludbrook, J. and Dudley, H., Why permutation tests are superior to *t* or *F* tests in biomedical research, *The American Statistician,* 52:127–32, 1998.
7. Neyman, J. and Pearson, E.S., On the use and interpretation of certain test criteria for purposes of statistical inference. Part I, *Biometrika,* 20A:175–240, 1928.
8. Lehmann, E.L., *Nonparametrics; Statistical Methods Based on Ranks*, 1st edition revised. Prentice Hall, Upper Saddle River, NJ, 1998.
9. Fisher, R.A., 'The coefficient of racial likeness' and the future of craniometry, *J. of the Royal Anthropological Soc.,* 66:57–63, 1936.
10. Kempthorne, O., The randomization theory of experimental inference, *J. of the Am. Stat. Assoc.,* 50:946–67, 1955.
11. Kempthorne, O. and Doerfler, T.E., The behavior of some significance tests under experimental randomization, *Biometrika,* 56:231–48, 1969.
12. Box, G.E.P. and Anderson, S.L., Permutation theory in the derivation of robust criteria and the study of departures from assumption, *J. of the Royal Stat. Soc.,* B 17:1–26, 1955.
13. Box, G.E.P., Hunter, W.G., and Hunter, J.S., *Statistics for Experimenters: An Introduction to Design, Data Analysis, and Model Building*, New York, Wiley, 1978.
14. Feinstein, A.R. Clinical biostatistics. XII. The role of randomization in sampling, testing, allocation, and credulous idolatry (Part 2), *J. of Clin. Pharmacol. and Ther.,* 14:898–915, 1973.
15. Edgington, E.S., *Randomization Tests*, 3rd edition, Marcel Dekker, New York, 1995.
16. Siegel, S. and Castellan, N.J., *Nonparametric Statistics for the Behavioral Sciences*, 2nd edition, McGraw-Hill, New York, 1988.
17. Agresti, A., *Categorical Data Analysis*, John Wiley and Sons, New York, 1990.
18. Davis, L.J., Exact tests for 2×2 contingency tables, *The Am. Stat.,* 40:139–41, 1986.
19. Pearson, K., On the criterion that a given system of deviations from the probable in the case of a correlated system of variables is such that it can reasonably be supposed to have arisen from random sampling, *Philosophical Magazine,* (5) 50, 157–75, 1900.
20. Fisher, R.A., On the interpretation of χ^2 from contingency tables and the calculation of P, *J. of the Royal Stat. Soc.,* 85:87–94, 1922.
21. Yates, F., Contingency tables involving small numbers and the χ^2 test, *J. of the Royal Stat. Soc.,* Suppl. 1, 217–35, 1934.
22. Cochran, W.G., Some methods for strengthening the common χ^2 tests, *Biometrics,* 10:417–51, 1954.
23. Sokal, R.R. and Rohlf, F.J., *Biometry*, 2nd edition, WH Freeman & Co., New York, 1981.
24. Freeman, G.H. and Halton, J.H., Note on exact treatment of contingency, goodness of fit and other problems of significance, *Biometrika,* 38:141–9, 1951.
25. Armitage, P., Test for linear trend in proportions and frequencies, *Biometrics,* 11:375–86, 1955.
26. Manly, B.F.J., *Randomization and Monte Carlo Methods in Biology*, Chapman & Hall, London, 1991.
27. Good, P., *Permutation Tests: A Practical Guide to Resampling Methods for Testing Hypotheses*, Springer-Verlag, New York, 1994.
28. Sprent, P., *Data Driven Statistical Methods*, Chapman & Hall, London, 1998.
29. Wilcoxon, F., Individual comparison by ranking methods, *Biometrics,* 1:80–3, 1945.

30. Mann, H.B. and Whitney, D.R., On a test of whether one of two random variables is stochastically larger than the other, *Ann. of Math. Stat.,* 18, 50–60, 1947.
31. Ludbrook, J., Repeated measurements and multiple comparisons in cardiovascular research, *Cardiovasc. Res.,* 28:303–11, 1994.
32. Neter, J., Wasserman, W., and Kutner, M.H., *Applied Linear Statistical Models*, 3rd edition, Richard D. Irwin, Homewood, IL, 1990.
33. Christopoulos, A., Assessing the distribution of parameters in models of ligand-receptor interaction: to log or not to log, *Trends in Pharmacological Science,* 19:351–7, 1998.
34. Morgan, B.J.T., *Elements of Simulation*, Chapman & Hall, London, 1986.
35. Everitt, B.S., *The Cambridge Dictionary of Statistics*, Cambridge University Press, Cambridge, 1998.
36. Ludbrook, J., Multiple comparison procedures updated, *Clin. and Exp. Pharmacol. and Physiol.,* 25:1032–7, 1998.
37. Concato, J. and Feinstein, A.R., Monte Carlo methods in clinical research: applications in multivariable analysis, *J. of Invest. Med.,* 45:394–400, 1997.
38. Lew, M.J. and Angus, J.A., Analysis of competitive agonist-antagonist interactions by nonlinear regression, *Trends in Pharmacological Science,* 16:328–37, 1995.
39. Efron, B. and Tibshirani, R.J., *An Introduction to the Bootstrap*, Chapman & Hall, New York, 1993.
40. Mooney, C.Z. and Duval, R.D., *Bootstrapping: A Nonparametric Approach to Statistical Inference*, Sage Publications, Newbury Park, 1993.
41. Ludbrook, J., Issues in biomedical statistics; comparing means by computer-intensive tests, *Australian and NZ Journal of Surgery,* 65:812–19, 1995.
42. Hall, P. and Martin, M., On the bootstrap and two-sample problems, *Australian Journal of Statistics,* 30A:179–92, 1988.
43. Fisher, N.I. and Hall, P., On bootstrap hypothesis testing, *Australian Journal of Statistics,* 32:177–90, 1990.
44. Fisher, N.I. and Hall, P., Bootstrap algorithms for small samples, *J. of Stat. Plann. and Inference,* 27:157–69, 1991.
45. Venables, W.N. and Ripley, B.D., *Modern Applied Statistics with S-PLUS*, 2nd edition, Springer, New York, 1997.

7 Beyond Eyeballing: Fitting Models to Experimental Data

Arthur Christopoulos and Michael J. Lew

CONTENTS

0-8493-0100-9/01/$0.00+$.50

7.1 INTRODUCTION

7.1.1 "EYEBALLING"

The oldest and most commonly used tool for examining the relationship between experimental variables is the graphical display. People are very good at recognizing patterns, and can intuitively detect various modes of behavior far more easily from a graph than from a table of numbers. The process of "eyeballing the data" thus represents the experimenter's first attempt at understanding their results and, in the past, has even formed the basis of formal quantitative conclusions. Eyeballing can sometimes be assisted by judicious application of a ruler, and often the utility of the ruler has been enhanced by linearizing data transformations. Nowadays it is more common to use a computer-based curve-fitting routine to obtain an "unbiased" analysis. In some common circumstances there is no important difference in the conclusions that would be obtained by the eye and by the computer, but there are important advantages of the more modern methods in many other circumstances. This chapter will discuss some of those methods, their advantages, and how to choose between them.

7.1.2 MODELS

The modern methods of data analysis frequently involve the fitting of mathematical models to the data. There are many reasons why a scientist might choose to model and many different conceptual types of models. Modeling experiments can be entirely constructed within a computer and used to test "what if" types of questions regarding the underlying mathematical aspects of the system of interest. In one sense, scientists are constructing and dealing with models all the time inasmuch as they form "worldview" models; experiments are designed and conducted and then used in an intuitive fashion to build a mental picture of what the data may be revealing about the experimental system (see Kenakin, this volume). The experimental results are then frequently analyzed by applying either empirical or mechanistic mathematical models to the data. It is these models that are the subject of this chapter.

7.2 EMPIRICAL OR MECHANISTIC?

Empirical models are simple descriptors of a phenomenon that serve to approximate the general shape of the relationship being investigated without any theoretical meaning being attached to the actual parameters of the model. In contrast, mechanistic models are primarily concerned with the quantitative properties of the relationship between the model parameters and its variables, that is, the processes that govern (or are thought to govern) the phenomenon of interest. Common examples of mechanistic models are those related to mass action that are applied to binding data to obtain estimates of chemical dissociation constants whereas nonmechanistic, empirical models might be any model applied to drug concentration–response curves in order to obtain estimates of drug potency. In general, mechanistic models are often the most useful, as they consist of a quantitative formulation of a hypothesis.[1] However, the consequences of using an inappropriate mechanistic model are worse than for empirical models because the parameters in mechanistic models provide information about the quantities and properties of real system components. Thus, the appropriateness of mechanistic models needs close scrutiny.

The designation of a mathematical model as either empirical or mechanistic is based predominantly on the purpose behind fitting the model to experimental data. As such, the *same* model can be both empirical and mechanistic depending on its context of use. As an example, consider the following form of the Hill equation:

$$Y = \frac{\alpha[A]^S}{[A]^S + K^S} \tag{1}$$

This equation is often used to analyze concentration–occupancy curves for the interaction of radioligands with receptors or concentration–response curves for the functional interaction of agonist drugs with receptors in cells or tissues. The Hill equation describes the observed experimental curve in terms of the concentration of drug (A), a maximal asymptote (α), a midpoint location (K), and a midpoint slope (S). In practice, these types of curves are most conveniently visualized on a semi-logarithmic scale, as shown in Figure 7.1.

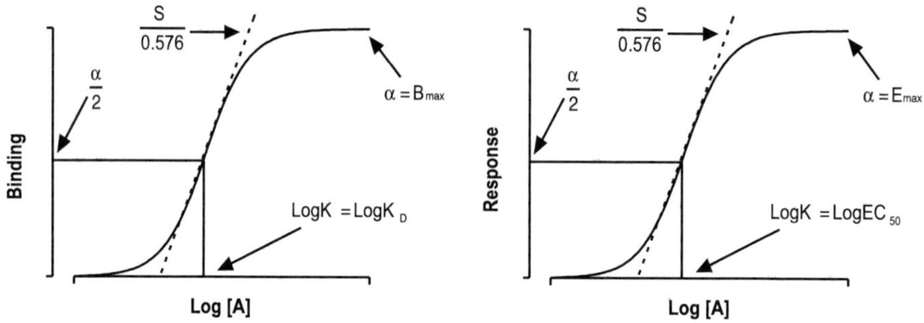

FIGURE 7.1 Concentration–binding (left) and concentration–response (right) curves showing the parameters of the Hill equation (α, K, and S) as mechanistic (left) or empirical (right) model descriptors.

When Hill first derived this equation,[2,3] he based it on a mechanistic model for the binding of oxygen to the enzyme, hemoglobin. In that context, the parameters that Hill was interested in, K and S, were meant to reveal specific biological properties about the interaction he was studying; K was a measure of the affinity of oxygen for the enzyme and S was the number of molecules of oxygen bound per enzyme. Subsequent experiments over the years have revealed that this model was inadequate in accounting for the true underlying molecular mechanism of oxygen-hemoglobin binding, but the equation remains popular both as a mechanistic model when its validity is accepted, and as an empirical model where its shape approximates that of experimental data. For instance, if the experimental curve is a result of the direct binding of a radioligand to a receptor, then application of Equation (1) to the dataset can be used to detect whether the interaction conforms to the simplest case of one-site mass-action binding and, if S = 1, the parameters K and α can be used as quantitative estimates of the ligand-receptor dissociation constant (K_D) and total density of receptors (B_{max}), respectively. This is an example where the Hill equation is a mechanistic equation, because the resulting parameters provide actual information about the underlying properties of the interaction. In contrast, concentration–response curves represent the final element in a series of sequential biochemical cascades that yield the observed response subsequent to the initial mass-action binding of a drug to its receptor. Thus, although the curve often retains a sigmoidal shape that is similar to the binding curve, the Hill equation is no longer valid as a mechanistic equation. Hence, the Hill equation is useful in providing a good fit to sigmoidal concentration–response curves, but the resulting parameters are considered empirical estimates of maximal response, midpoint slope, and midpoint location, and no mechanistic interpretation should be made.

7.3 TYPES OF FITTING

The variables whose relationships that can be plotted on Cartesian axes do not necessarily have the same properties. Often one variable is controlled by the experimenter and the other variable is a measurement. Thus one variable has substantially more uncertainty or variability than the other, and traditionally that variable would be plotted on the vertical axis. In that circumstance the Y variable can be called the "dependent" variable because of its dependence on the underlying relationship and on the other variable, which is called "independent" to denote its higher reliability. It is important to note that not all datasets have a clearly independent variable. Historically, the statistical determination of the relationship between two or more dependent variables has been referred to as a *correlation analysis*, whereas the determination of the relationship between dependent and independent variables has come to be known as a *regression analysis*. Both types of analyses, however, can share a number of common features, and some are discussed below.

7.3.1 Correlation

Correlation is not strictly a regression procedure, but in practice it is often confused with linear regression. Correlation quantifies the degree by which two variables vary together. It is meaningful only when both variables are outcomes of measurement such that there is no independent variable.

7.3.1.1 The Difference Between Correlation and Linear Regression

Correlation quantifies how well two dependent variables vary together; linear regression finds the line that best predicts a dependent variable given one or more independent variables, that is, the "line of best-fit."[4] Correlation calculations do not find a best-fit straight line.[5]

7.3.1.2 The Meaning of r^2

The *direction* and *magnitude* of the correlation between two variables can be quantified by the correlation coefficient, r, whose values can range from –1 for a perfect negative correlation to 1 for a perfect positive correlation. A value of 0, of course, indicates a lack of correlation. In interpreting the meaning of r, a difficulty can arise with values that are somewhere between 0 and –1 or 0 and 1. Either the variables do influence each other to some extent, or they are under the influence of an additional factor or variable that was not accounted for in the experiment and analysis. A better "feel" for the covariation between two variables may be derived by squaring the value of the correlation coefficient to yield the *coefficient of determination,* or r^2 value. This number may be defined as the fraction of the variance in the two variables that is shared, or the fraction of the variance in one variable that is explained by the other (provided the following assumptions are valid). The value of r^2, of course, will always be between 0 and 1.

7.3.1.3 Assumptions of Correlation Analysis

1. The subjects are randomly selected from a larger population. This is often not true in biomedical research, where randomization is more common

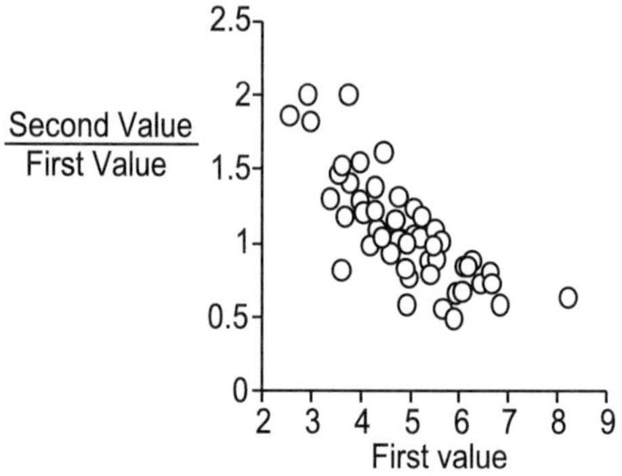

FIGURE 7.2 An apparent correlation between two sets of unrelated random numbers (pseudo-random numbers generated with mean = 5 and standard deviation = 1) comes about where the Y value is expressed as a function of the X value (here each Y value is expressed as a fraction of the corresponding X value).

than sampling, but may be sufficient to assume that the subjects are at least representative of a larger population.

2. The samples are paired, i.e., each experimental unit has both X and Y values.
3. The observations are independent of each other. Sampling one member of the population should not affect the probability of sampling another member (e.g., making measurements in the same subject twice and treating them as separate datapoints; making measurements in siblings).
4. The measurements are independent. If X is somehow involved or connected to the determination of Y, or vice versa, then correlation is not valid. This assumption is very important because artifactual correlations can result from its violation. A common cause of such a problem is where the Y value is expressed as either a change from the X value, or as a fraction of the corresponding X value (Figure 7.2).
5. The X values were measurements, not controlled (e.g., concentration, etc.). The confidence interval for r^2 is otherwise meaningless, and we must then use linear regression.
6. The X and Y values follow a Gaussian distribution.
7. The covariation is linear.

7.3.1.4 Misuses of Correlation Analysis

Often, biomedical investigators are interested in comparing one method for measuring a biological response with another. This usually involves graphing the results as an X, Y plot, but what to do next? It is quite common to see a correlation analysis applied to the two methods of measurement and the correlation coefficient, r, and the resulting *P* value utilized in hypothesis testing. However, Ludbrook[6] has outlined some serious criticisms of this approach, the major one being that although correlation analysis will identify the strength of the linear association between X and Y, as it is intended to do, it will give no indication of any bias between the two methods of measurement. When the purpose of the exercise is to identify and quantify fixed and proportional biases between two methods of measurement, then correlation analysis is inappropriate, and a technique such as ordinary or weighted least products regression[6] should be used.

7.3.2 Regression

The actual term "regression" is derived from the latin word "*regredi*," and means "to go back to" or "to retreat." Thus, the term has come to be associated with those instances where one "retreats" or "resorts" to approximating a response variable with an estimated variable based on a functional relationship between the estimated variable and one or more input variables. In regression analysis, the input (independent) variables can also be referred to as "regressor" or "predictor" variables.

7.3.2.1 Linear Regression

The most straightforward methods for fitting a model to experimental data are those of linear regression. Linear regression involves specification of a linear relationship between the dependent variable(s) and certain properties of the system under investigation. Surprisingly though, linear regression deals with some curves (i.e., non-straight lines) as well as straight lines, with regression of straight lines being in the category of "ordinary linear regression" and curves in the category of "multiple linear regressions" or "polynomial regressions."

7.3.2.2 Ordinary Linear Regression

The simplest general model for a straight line includes a parameter that allows for inexact fits: an "error parameter" which we will denote as ε. Thus we have the formula:

$$Y = \alpha + \beta X + \varepsilon \tag{2}$$

The parameter, α, is a *constant*, often called the "intercept" while β is referred to as a *regression coefficient* that corresponds to the "slope" of the line. The additional parameter, ε, accounts for the type of error that is due to random variation caused by experimental imprecision, or simple fluctuations in the state of the system from one time point to another. This error term is sometimes referred to as the *stochastic* component of the model, to differentiate it from the other, *deterministic*, component of the model (Figure 7.3).[7] When data are fitted to the actual straight-line model, the error term denoted by ε is usually not included in the fitting procedure so that the output of the regression forms a perfect straight line based solely on the deterministic component of the model. Nevertheless, the regression procedure assumes that the scatter of the datapoints about the best-fit straight line reflects the effects of the error term, and it is also implicitly assumed that ε follows a Gaussian distribution with a mean of 0. This assumption is often violated, however, and the implications

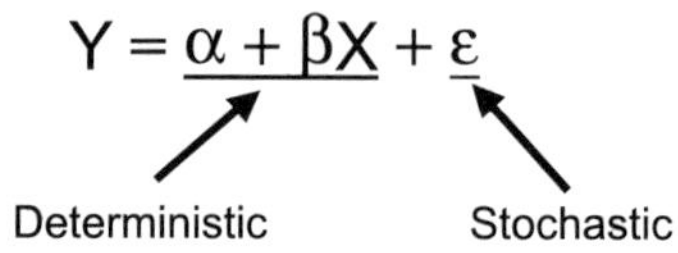

FIGURE 7.3 The simple linear population model equation indicating the deterministic component of the model that is precisely determined by the parameters α and β, and the stochastic component of the model, ε, that represents the contribution of random error to each determined value of Y.

are discussed elsewhere in this chapter. For now, however, we will assume that the error is Gaussian; Figure 7.4 illustrates the output of the linear model with the inclusion of the error term. Note that the Y values of the resulting "line" are randomly distributed above and below the ideal (dashed) population line defined by the deterministic component of the model.

7.3.2.3 Multiple Linear Regression

The straight line equation [Equation (2)] is the simplest form of the linear regression model, because it only includes one independent variable. When the relationship of interest can be described in terms of more than one independent variable, the regression is then defined as "multiple linear regression." The general form of the linear regression model may thus be written as:

$$Y = \alpha + \beta_1 X_1 + \beta_2 X_2 + \ldots + \beta_i X_i + \varepsilon \tag{3}$$

where Y is the dependent variable, and X_1, X_2 … X_i are the (multiple) independent variables. The output of this model can deviate from a straight line, and one may thus question the meaning of the word "linear" in "linear regression." Linear regression implies a linear relationship between the dependent variable and the *parameters*, not the independent variables of the model. Thus Equation (3) is a linear model

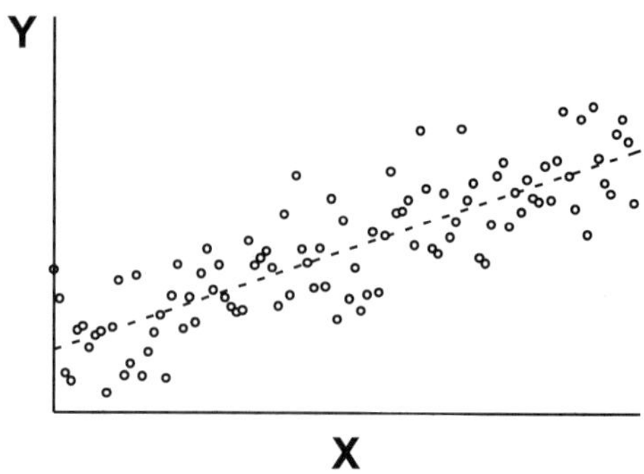

FIGURE 7.4 A linear model that incorporates a stochastic (random error) component. The dashed line is the deterministic component, whereas the points represent the effect of random error [denoted by the symbol ε in Equation (2)].

because the parameters α, β_1, β_2 … β_i have the (implied) exponent of unity. Multiple linear regression models also encompass *polynomial functions*:

$$Y = \alpha + \beta_1 X + \beta_2 X^2 + \ldots + \beta_i X^i + \varepsilon \tag{4}$$

The equation for a straight line [Equation (2)] is a first-order polynomial. The quadratic equation, $Y = \alpha + \beta_1 X + \beta_2 X^2$, is a second-order polynomial whereas the cubic equation, $Y = \alpha + \beta_1 X + \beta_2 X^2 + \beta_3 X^3$ is a third-order polynomial. Each of these higher order polynomial equations defines curves, not straight lines. Mathematically, a linear model can be identified by taking the first derivative of its deterministic component with respect to the parameters of the model. The resulting derivatives should not include any of the parameters; otherwise, the model is said to be "nonlinear." Consider the following second-order polynomial model:

$$Y = \alpha + \beta_1 X + \beta_2 X^2 \tag{5}$$

Taking first derivatives with respect to each of the parameters yields:

$$\frac{\partial Y}{\partial \alpha} = 1 \tag{6}$$

$$\frac{\partial Y}{\partial \beta_1} = X \tag{7}$$

$$\frac{\partial Y}{\partial \beta_2} = X^2 \tag{8}$$

The model is linear because the first derivatives do not include the parameters. As a consequence, taking the second (or higher) order derivative of a linear function with respect to its parameters will always yield a value of zero.[8] Thus, if the independent variables and all but one parameter are held constant, the relationship between the dependent variable and the remaining parameter will always be linear.

It is important to note that linear regression does not actually test whether the data sampled from the population follow a linear relationship. It assumes linearity and attempts to find the best-fit straight line relationship based on the data sample.

7.3.2.4 Nonlinear Regression

Because there are so many types of nonlinear relationships, a general model that encompasses all their behaviors cannot be defined in the sense used above for linear models, so we will define an explicit nonlinear function for illustrative purposes. In this case, we will use the Hill equation [Equation (1); Figure 7.1] which contains one independent variable [A], and 3 parameters, α, K, and S. Differentiating Y with respect to each model parameter yields the following:

$$\frac{\partial Y}{\partial \alpha} = \frac{[A]^S}{[A]^S + K^S} \tag{9}$$

$$\frac{\partial Y}{\partial K} = \frac{-\alpha S(K[A])^S}{K([A]^S + K^S)^2} \tag{10}$$

$$\frac{\partial Y}{\partial S} = \frac{-\alpha(K[A])^S(\ln K - \ln[A])}{([A]^S + K^S)^2} \tag{11}$$

All derivatives involve at least two of the parameters, so the model is nonlinear. However, it can be seen that the partial derivative in Equation (9) does not contain the parameter, α. A linear regression of Y on $[A]^S/(K^S + [A]^S)$ will thus allow the estimation of α. Because this last (linear) regression is conditional on knowing the values of K and S, α is referred to as a "conditionally linear" parameter. Nonlinear models that contain conditionally linear parameters have some advantages when it comes to actual curve fitting.[7]

7.3.2.5 Assumptions of Standard Regression Analyses[4,7]

1. The subjects are randomly selected from a larger population. The same caveats apply here as with correlation analyses.
2. The observations are independent.
3. X and Y are not interchangeable. Regression models used in the vast majority of cases attempt to predict the dependent variable, Y, from the independent variable, X and assume that the error in X is negligible. In special cases where this is not the case, extensions of the standard regression techniques have been developed to account for nonnegligible error in X.
4. The relationship between X and Y is of the correct form, i.e., the expectation function (linear or nonlinear model) is appropriate to the data being fitted.
5. The variability of values around the line is Gaussian.
6. The values of Y have constant variance. Assumptions 5 and 6 are often violated (most particularly when the data has variance where the standard deviation increases with the mean) and have to be specifically accounted for in modifications of the standard regression procedures.
7. There are enough datapoints to provide a good sampling of the random error associated with the experimental observations. In general, the minimum number of independent points can be no less than the number of parameters being estimated, and should ideally be significantly higher.

7.4 HOW IT WORKS

7.4.1 Minimizing an Error Function (Merit Function)

The goal of both linear and nonlinear regression procedures is to derive the "best fit" of a particular model to a set of experimental observations. To obtain the best-fit curve we have to find parameter values that minimize the difference between the observed experimental observations and the chosen model. This difference is assumed to be due to the error in the experimental determination of the datapoints, and thus it is common to see the entire model-fitting process described in terms of "minimization of an error function" or minimization of a "merit function."[9]

The most common representation ("norm") of the merit function for regression models is based on the chi-square distribution. This distribution and its associated statistic, χ^2, have long been used in the statistical arena to assess "goodness-of-fit" with respect to identity between observed and expected frequencies of measures. Because regression analyses also involve the determination of the best model estimates of the dependent variables based on the experimentally observed dependent variables, it is quite common to see the function used to determine the best-fit of the model parameters to the experimental data referred to as the "χ^2 function," and the procedure referred to as "chi-square fitting."[9]

7.4.2 Least Squares

The most widely used method of parameter estimation from curve fitting is the method of least squares. To explain the principle behind least squares methods, we will use an example, in this case the simple linear model. Theoretically, finding the slope, β, and intercept, α, parameters for a perfect straight line is easy: any two X,Y pairs of points can be utilized in the familiar "rise-over-run" formulation to obtain the slope parameter, which can then be inserted into the equation for the straight line to derive the intercept parameter. In reality, however, experimental observations that follow linear relationships almost never fall exactly on a straight line due to random error. The task of finding the parameters describing the line is thus no longer simple; in fact, it is unlikely that values for α and β defined by any pair of experimental points will describe the best line through all the points. This is illustrated in Figure 7.5; although the dataset appears to follow a linear relationship, it can be seen that different straight lines, each characterized by different slopes and intercepts, are derived depending on which two X,Y pairs are used.

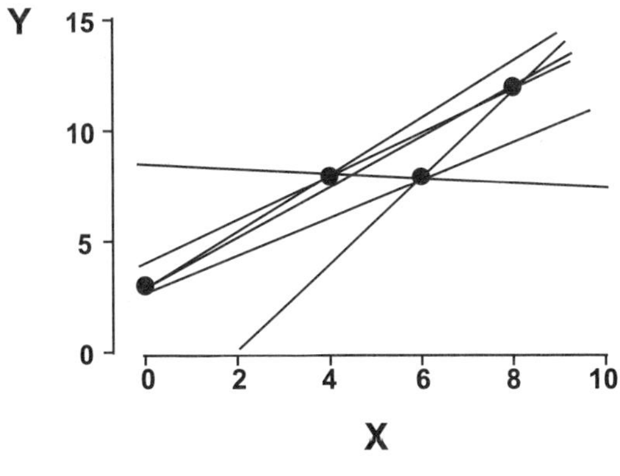

FIGURE 7.5 All possible straight lines that can be drawn through a four-point dataset when only two points are used to define each line.

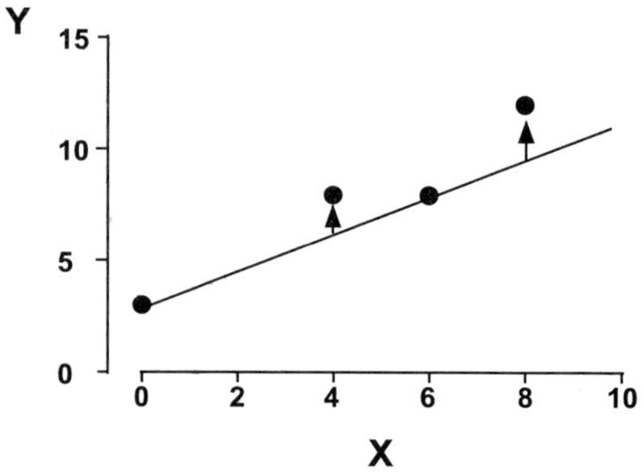

FIGURE 7.6 A combination of zero and nonzero residuals. The dataset is the same as in Figure 7.5, with only one of the lines now drawn through the points. The vertical distance of each point from the line (indicated by the arrows) is defined as the "residual."

What is needed, therefore, is a "compromise" method for obtaining an objective best-fit. We begin with our population model [Equation (2)]:

$$Y = \alpha + \beta X + \varepsilon$$

and derive an equation that is of the same form:

$$\hat{Y} = \hat{\alpha} + \hat{\beta} X \tag{12}$$

where $\hat{Y}$ is the *predicted* response and $\hat{\alpha}$ and $\hat{\beta}$ are the *estimates* of the population intercept and slope parameters, respectively. The difference between the response variable, Y, and its predictor, $\hat{Y}$, is called the "*residual*" and its magnitude is therefore a measure of how well $\hat{Y}$ predicts Y. The closer the residual is to a value of zero for each experimental point, the closer the predicted line will be to that point. However, because of the error in the data (the ε term in the population model), no prediction equation will fit all the datapoints exactly and, hence, no equation can make the residuals all equal zero. In the example above, each straight line will yield a residual of zero for two points, but a nonzero residual for the other two points; Figure 7.6 illustrates this for one of the lines.

A best-fit compromise is found by *minimizing the sum of the squares of the residuals*, hence the name "least squares." Mathematically, the appropriate merit function can be written as:

$$\chi^2 = \sum_{i=1}^{N} \left(\frac{Y_i - f(X_i, \theta)}{w_i} \right)^2 = \sum_{i=1}^{N} \left(\frac{r_i}{w_i} \right)^2 \tag{13}$$

where χ^2 is the weighted sum of the squares of the residuals (r_i) and is a function of the parameters (the vector, θ), and the N datapoint, X_iY_i. The term, w_i, is the statistical

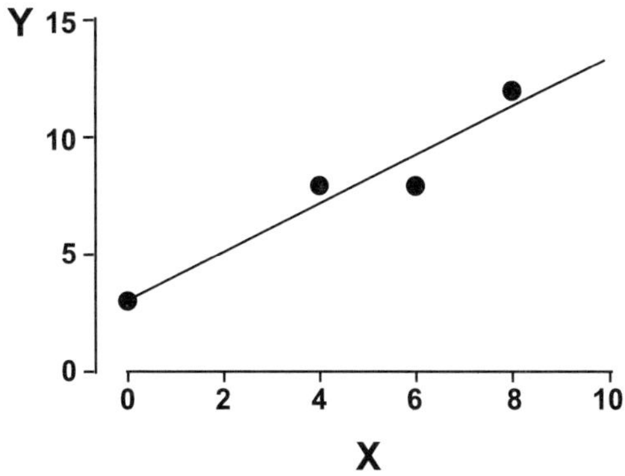

FIGURE 7.7 The minimized least squares fit of the straight line model [Equation (2)] to the dataset shown in Figures 7.5 and 7.6.

weight (see below) of a particular datapoint, and when used, most often relates to the standard error of that point. For standard (unweighted) least squares procedures such as the current example, w_i equals 1. The least squares fit of the dataset outlined above is shown in Figure 7.7. Note that the best-fit straight line yields nonzero residuals for three of the four datapoints. Nevertheless, the resulting line is based on parameter estimates that give the smallest sum-of-squares of those residuals.

Why do we use the sum of the square of the residuals and not another norm of the deviation, such as the average of the absolute values of the residuals? Arguably, simply because of convention! Different norms of deviation have different relative sensitivities to small and large deviations and conventional usage suggests that sums of the square residuals represent a sensible compromise.[4,10] The popularity of least squares estimators may also be based on the fact that they are relatively easy to determine and that they are accurate estimators if certain assumptions are met regarding the independence of errors and a Gaussian distribution of errors in the data.[8,9,11] Nonetheless, for extremely large deviations due to outlier points, least squares procedures can fail in providing a sensible fit of the model to the data.

Although the example used above was based on a linear model, nonlinear least squares follow the same principles as linear least squares and are based on the same assumptions. The main difference is that the sum-of-squares merit function for linear models is well-behaved and can be solved analytically in one step, whereas for nonlinear models, iterative or numerical procedures must be used instead.

In most common applications of the least squares method to linear and nonlinear models, it is assumed that the majority of the error lies in the dependent variable. However, there can be circumstances when both X and Y values are attended by random error, and different fitting approaches are warranted. One such approach has been described by Johnson,[12] and is particularly useful for fitting data to nonlinear models. In essence, Johnson's method utilizes a form of the standard χ^2 merit function, given above, that has been expanded to include the "best-fit" X value and its associated variance. The resulting merit function is then minimized using an appropriate least squares curve fitting algorithm.

7.4.3 Nonleast Squares

Cornish-Bowden[11] has listed the minimal requirements for optimal behavior of the least squares method:

a. Correct choice of model.
b. Correct data weighting is known.
c. Errors in the observations are independent of one another.
d. Errors in the observations are normally distributed.
e. Errors in the observations are unbiased (have zero mean).

And we can add:

f. None or the datapoints are erroneous (outliers).

Often, however, the requirements for optimal behavior cannot be met. Other techniques are available for deriving parameter estimates under these circumstances, and they are generally referred to as "robust estimation" or "robust regression" techniques. Because the word "robustness" has a particular connotation, it is perhaps unfair to class all of the diverse nonleast squares procedures under the same umbrella. Overall, however, the idea behind robust estimators is that they are more insensitive to deviations from the assumptions that underlie the fitting procedure than least squares estimators.

"Maximum likelihood" calculations are one class of robust regression techniques that are not based on a Gaussian distribution of errors. In essence, regression procedures attempt to find a set of model parameters that generate a curve that best matches the observed data. However, there is no way of knowing which parameter set is the correct one based on the (sampled) data, and thus there is no way of calculating a probability for any set of fitted parameters being the "correct set." Maximum likelihood calculations work in the opposite direction, that is, given a particular model with a particular set of parameters, maximum likelihood calculations derive a probability for the *data* being obtained. This (calculated) probability of the data, given the parameters, can also be considered to be the likelihood of the parameters, given the data.[9] The goal is then to fit for a set of parameters that maximize this likelihood, hence the term "maximum likelihood," and the calculations attempt to find the regression that has the maximum likelihood of producing the observed dataset. It has been pointed out that there is no formal mathematical basis for the maximum likelihood procedure and because maximum likelihood calculations are quite involved, they are not routinely utilized explicitly.[9] Fortunately the simpler least squares methods described above are equivalent to maximum likelihood calculations where the assumptions of linear and nonlinear regression (particularly the independence and Gaussian distribution of the errors in the data) are valid.[8,9,11]

Certain robust regression techniques focus on using measures of central tendency other than the mean as the preferred statistical parameter estimator. For instance, Cornish-Bowden[11] has described how the median is more insensitive to outlier points in linear regression and certain cases of nonlinear regression than the mean. A drawback of this approach, however, is that it quickly becomes cumbersome when extended to more complex linear problems.

7.4.4 Weighting

The simplest minimization functions make no distinction between different experimental points, and assume that each observation contributes equally to the estimation of model parameters. This is appropriate when the variance of all the observations is uniform, and the error is referred to as *homoscedastic*. However, in reality it is common that different points have different variances associated with them with the result that the points with the most variance may have an undue influence on the parameters obtained from an unweighted curve fit. For example, results from many biological experiments are often expressed as a change from a baseline value, with the consequence that the points near the baseline become small numbers (near zero) with a low variance. Points representing larger responses will naturally have a larger variance, a situation that can be described as *heteroscedasticity*. An unweighted curve fit through heteroscedastic data will allow the resulting curve to deviate from the well-defined (tight) near-zero values to improve the fit of the larger, less well-defined values. Clearly it would be better to have the fit place more credence in the more reliably estimated points, something that can be achieved in a weighted curve fit.

Equation (13) was used previously to define the general, least squares, minimization function. There are a number of variations available for this function that employ differential data weighting.[13] These functions explicitly define a value for the w_i term in Equation (13). For instance, if $w_i = 1$ or a constant, then the weighting is said to be "uniform"; if $w_i = Y_i$, then

$$\chi^2 = \sum_{i=1}^{N}\left(\frac{r_i}{Y_i}\right)^2 = \sum_{i=1}^{N}\frac{1}{Y_i^2}(r_i)^2$$

and the weighting is said to be "relative." Relative weighting is also referred to as "weighting by $1/Y^2$" and is useful where the experimental uncertainty is a constant fraction of Y. For example, counts of radioactive decay will have variances described by the Poisson distribution where the variance scales with the mean, and thus the likely error in each estimate is a constant percentage of counts rather than a constant value for any number of counts. Thus, a curve fit allowing for relative weighting can adjust for the resulting heteroscedastic variance. Another useful weighting value is $w_i = \sqrt{Y_i}$. This yields "weighting by $1/Y$" and is appropriate, for example, when most of the experimental uncertainty in the dependent variable is due to some sort of counting error.[5] Other weighting schemes utilize the number of replicates that are measured for each value of Y to determine the appropriate weight for the datapoints.[13]

7.4.5 Regression Algorithms

What are the actual "mechanics" that underlie the χ^2 minimization process behind least squares regression techniques? The χ^2 merit function for linear models (including polynomials) is quadratic in nature, and is thus amenable to an exact analytical solution. In contrast, nonlinear problems must be solved iteratively, and this procedure can be summarized as follows:

a. Define the merit function.
b. Start with a set of initial estimates (guesses) of the regression parameters and determine the value of the merit function for this set of estimates.
c. Adjust the parameter estimates and recalculate the merit function. If the merit function is improved, then keep the parameter values as new estimates.
d. Repeat step c (each repeat is an "iteration"). When further iterations yield a negligible improvement in the fit, stop adjusting the parameter estimates and generate the curve based on the last set of estimates.

The rules for adjusting the parameters of the nonlinear model are based on matrix algebra and are formulated as computer algorithms. The merit function can be viewed as a multidimensional surface that has all possible sum-of-squares values as one plane and all possible values of each of the model parameters as the other planes. This surface may thus vary from a smooth, symmetrical shape to one characterized by many crests and troughs. The role of the nonlinear regression algorithm is to work its way down this surface to the deepest trough that should then correspond to the set of model parameters that yield the minimum sum-of-squares value.

There are a number of different algorithms that have been developed over the years, and they all have their pros and cons. One of the earliest algorithms is the *method of steepest descent* (or the *gradient search method*[8]). This method proceeds down the steepest part of the multidimensional merit function surface in fixed step lengths that tend to be rather small.[9] At the end of each iteration, a new slope is calculated and the procedure repeated. Many iterations are required before the algorithm converges on a stable set of parameter values. This method works well in the initial iterations, but tends to drag as it approaches a minimum value.[13]

The *Gauss-Newton method* is another algorithm that relies on a linear approximation of the merit function. By making this approximation, the merit function approaches a quadratic, its surface becomes a symmetrical ellipsoid, and the iterations of the Gauss-Newton algorithm allow it to converge toward a minimum much more rapidly than the method of steepest descent. The Gauss-Newton method works best when it is employed close to the surface minimum, because at this point most merit functions are well approximated by linear (e.g., quadratic) functions.[9] In contrast, the Gauss-Newton method can work poorly in initial iterations, where the likelihood of finding a linear approximation to the merit function is decreased.

A method exploiting the best features of the methods of steepest descent and Gauss-Newton was described by Marquardt, based on an earlier suggestion by Levenberg,[9] and the resulting algorithm is thus often referred to as the *Levenberg-Marquardt method*. Marquardt realized that the size of the increments in an interative procedure poses a significant scaling problem for any algorithm, and proceeded to refine the scaling issue and derive a series of equations that can approximate the steepest descent method at early iterations and the Gauss-Newton method at later stages closer to the minimum. The Levenberg-Marquard method (sometimes simply referred to as the Marquardt method) has become one of the most widespread algorithms used for computerized nonlinear regression.

Another type of algorithm that is geometric rather than numeric in nature is the *Nelder-Mead Variable Size Simplex method*.[8,14] Unlike the methods outlined above, this method does not require the calculation of any derivatives. Instead, this algorithm depends on the generation of a number of starting points, called "vertices," based on initial estimates for each parameter of the model, as well as an initial increment step. The vertices form a multidimensional shape called a "simplex." The goodness of fit is evaluated at each vertex in the simplex, the worst vertex is rejected and a new one is generated by combining desirable features of the remaining vertices. This is repeated in an iterative fashion until the simplex converges to a minimum. The big advantage of the Nelder-Mead method is that it is very successful in converging to a minimum; its main disadvantage is that it does not provide any information regarding the errors associated with the final parameter estimates.[8]

7.5 WHEN TO DO IT (APPLICATION OF CURVE FITTING PROCEDURES)

7.5.1 Calibration Curves (Standard Curves)

Calibration curves are most convenient when they are linear, but even for assays where a linear relationship is expected on theoretical grounds, nonlinear curves can result from instrumentation nonlinearities and other factors. The equation of a curve fitted through the calibration data will allow convenient conversion between the raw measurement and the required value. In cases where there is no theoretical basis for choosing one model over another, calibration curves can be considered to be a smoothing rather than a real fitting problem and one might decide to apply a polynomial model to the data because of the availability of an analytical solution. In such a case the order of the chosen polynomial would need to be low so that noise in the calibration measurements is not converted into wobbles on the calibration curve.

7.5.2 Parameterization of Data (Distillation)

It is often desirable to describe data in an abbreviated way. An example of this is the need to summarize a concentration–response curve into a potency estimate and maximum response value. These parameters are easily obtained by eyeballing the data, but an unbiased estimate from an empirical curve fit is preferable and probably more acceptable to referees!

7.6 HOW TO DO IT

7.6.1 Choosing the Right Model

7.6.1.1 Number of Parameters

The expectation function should include the *minimum* number of parameters that adequately define the model and that allow for a successful convergence of the fit.

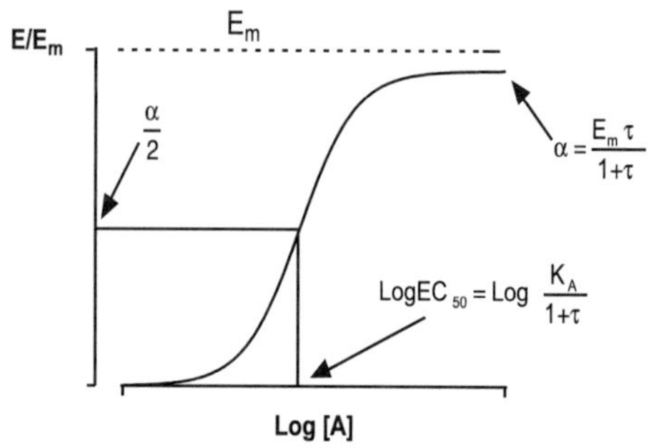

FIGURE 7.8 The relationship between the Hill equation [Equation (15)] parameters, α and EC_{50}, and the operational model [Equation (14)] parameters K_A, τ, and E_m, in the description of a concentration–response curve of an agonist drug. It can be seen that each parameter of the Hill equation is composed of two operational model parameters.

If a model is overparameterized, it is considered to possess "redundant" parameters (often used interchangeably with the term "redundant variables"), and the regression procedure will either fail or yield meaningless parameter estimates. Consider the "operational model" of Black and Leff.[15] This is a model that is often used in pharmacological analyses to describe the concentration–response relationship of an agonist (A) in terms of its affinity (dissociation constant) for its receptor (K_A), its "operational" efficacy (τ), and the maximum response (E_m) that the tissue can elicit. One common form of the model is:

$$E = \frac{E_m \cdot \tau \cdot [A]}{([A] + K_A) + \tau \cdot [A]} \tag{14}$$

where E denotes the observed effect. Figure 7.8 shows a theoretical concentration–response curve, plotted in semilogarithmic space, that illustrates the relationship between the operational model parameters and the maximal asymptote (α) and midpoint location (EC_{50}) of the resulting sigmoidal curve. A concentration–response curve like the one in Figure 7.8 can be successfully fitted using the two-parameter version of the Hill equation, which describes the curve in terms of only the EC_{50} and α (the slope being equal to 1):

$$E = \frac{\alpha \cdot [A]}{[A] + EC_{50}} \tag{15}$$

However, it can be seen in Figure 7.8 that the midpoint and maximal asymptote of the curve are related to the operational model in a more complicated manner; each parameter of the sigmoidal Hill equation is comprised of two operational model parameters. If someone were to try directly fitting Equation (14) to this curve in order to derive individual estimates of E_m, K_A, and τ, they would be unsuccessful. As it stands, the operational model is overparameterized for fitting to a single curve;

the regression algorithm simply will not be able to apportion meaningful estimates between the individual operational model parameters as it tries to define the midpoint and maximal asymptote of the concentration–response curve. In practice, the successful application of the operational model to real datasets requires additional experiments to be incorporated in the curve fitting process that allow for a better definition of the individual model parameters.[16,17]

7.6.1.2 Shape

When fitting empirical models to data the most important feature of the model must be that its shape should be similar to the data. This seems extraordinarily obvious, but very little exploration of the literature is needed to find examples where the curve and the data have disparate shapes! Empiricism allows one a great deal of freedom in choosing models, and experimenters should not be overly shy of moving away from the most common models (e.g., the Hill equation) when their data ask for it. Even for mechanistic models it is important to look for a clear shape match between the model and data: a marked difference can only mean that the model is inappropriate or the data of poor quality.

Perhaps the only feature that practically all biological responses have in common is that they can be approximated by nonlinear, saturating functions. When plotted on a logarithmic concentration scale, responses usually lie on a sigmoid curve, as shown in Figures 7.1 and 7.8, and a number of functions have been used in the past to approximate the general shape of such responses. Parker and Waud,[18] for instance, have highlighted that the rectangular hyperbola, the integral of the Gaussian distribution curve, the arc-tangent, and the logistic function have all been used by various researchers to empirically fit concentration–response data. Some of these functions are more flexible than others; for instance, the rectangular hyperbola has a fixed slope of 1. In contrast, the logistic equation has proven very popular in the fitting of concentration–response data:

$$E = \frac{1}{1 + e^{-(\alpha + \beta X)}} \tag{16}$$

Part of the popularity of this equation is its flexibility and its ability to match the parameters of the Hill equation [Equation (1)] for empirical fitting purposes.

In general, the correct choice of expectation function is most crucial when fitting mechanistic models. The difficulty in ascertaining the validity of the underlying model in these cases arises because the curve fitting process is undertaken with the automatic assumption that the model is a plausible one prior to actually fitting the model and applying some sort of diagnostics to the fit (see Assessing the Quality of the Fit, below). We must always remain aware, therefore, that we will never really know the "true" model, but can at least employ a reasonable one that accommodates the experimental findings and, importantly, allows for the prediction of testable hypotheses. From a practical standpoint, this may be seen as having chosen the "right" mechanistic model.

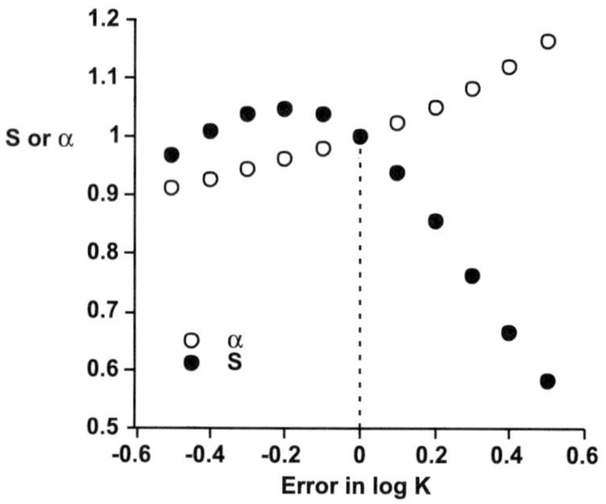

FIGURE 7.9 Altered estimates of the maximal asymptote, α, and the slope, S, obtained by fitting the Hill equation to logistic data where the parameter K (log K) was constrained to differ from the correct value. The systematic relationship between the error in K and the values of the parameters S and α indicates that each is able to partially correct for error in K and thus are correlated with K.

7.6.1.3 Correlation of Parameters

When a model, either mechanistic or empirical, is applied to a dataset we generally consider each of the parameters to be responsible for a single property of the curve. Thus, in the Hill equation, there is a slope parameter, S, a parameter for the maximum asymptote (α), and a parameter for the location (K or EC_{50}). Ideally, each of these parameters would be entirely independent so that error or variance in one does not affect the values of the others. Such a situation would mean that the parameters are entirely uncorrelated. In practice it is not possible to have uncorrelated parameters (see Figure 7.9), but the parameters of some functions are less correlated than others. Strong correlations between parameters reduce the reliability of their estimation as well as making any estimates from the fit of their variances overly optimistic.[19]

7.6.1.4 Distribution of Parameters

The operational model example can also be used to illustrate another practical consideration when entering equations for curve fitting, namely the concept of "reparameterization."[13] When fitting the operational model or the Hill equation to concentration–response curves, the parameters may be entered in the equation in a number of ways; for instance, the EC_{50} is commonly entered as $10^{LogEC_{50}}$. This reparameterization means that the regression algorithm will actually provide the best-fit estimate of the logarithm of the EC_{50}. Why reparameterize? As mentioned earlier, many of the assumptions of nonlinear regression rely on a Gaussian distribution of experimental uncertainties. Many model parameters, including the EC_{50} of the Hill equation, the dissociation constant of a hyperbolic radioligand binding equation, and the τ parameter of the operational model, follow an approximately Gaussian distribution only when transformed into logarithms.[17] Thus, although not particularly important for the estimation of the parametric value, reparameterization can improve the validity of statistical inferences made from nonlinear regression algorithms.[13] Other examples of reparameterizations that can increase the statistical reliability of the estimation procedure include recasting time parameters as reciprocals and counts of radioactive decay as square roots.[5]

7.6.2 Assessing the Quality of the Fit

The final determination of how "appropriate" the fit of a dataset is to a model will always depend on a number of factors, including the degree of rigor the researcher actually requires. Curve fitting for the determination of standard curves, for instance, will not warrant the same diagnostic criteria one may apply to a curve fit of an experimental dataset that was designed to investigate a specific biological mechanism. In the case of standard curves, an eyeball inspection of the curve superimposed on the data is usually sufficient to indicate the reliability of the fit for that specific purpose. However, when the fitting of models to experimental data is used to provide insight into underlying biological mechanisms, the ability to ascribe a high degree of appropriateness to the resulting curve fit becomes paramount.

7.6.2.1 Inspection

Although usually sufficient for empirical models, an initial test for conformity of the data to any selected model is a simple inspection of the curve fit superimposed on the data. Although rudimentary, this procedure is quite useful in highlighting *really bad* curve fits, i.e., those that are almost invariably the consequence of having inadvertently entered the wrong equation or setting certain parameter values to a constant value when they should have been allowed to vary as part of the fitting process. Assuming that visual inspection does not indicate a glaring inconsistency of the model with the data, there are a number of statistical procedures that can be used to quantify the goodness of the fit.

7.6.2.2 Root Mean Square

Figure 7.10 shows a schematic of an experimental dataset consisting of 6 observations (open circles labeled obs1 – obs6) and the superimposed best-fit of a sigmoidal concentration–response model [Equation (15)] to the data. The solid circles (exp1 – exp6) represent the expected response corresponding to each X-value used for the determination of obs1 – obs6, derived from the model fit. The sum of the squared residuals, i.e., the sum of the squared differences between the observed and expected responses has also been defined as the *Error Sum of Squares* (SSE), and it is this quantity that most researchers think of when discussing the sum-of-squares derived from their curve fitting exercises [see Equation (13)]:

$$\text{SSE} = (\text{obs1} - \text{exp1})^2 + (\text{obs2} - \text{exp2})^2 + \ldots + (\text{obs6} - \text{exp6})^2 \tag{17}$$

The SSE is sometimes used as an index of goodness-of-fit; the smaller the value, the better the fit. However, in order to use this quantity more effectively, an allowance

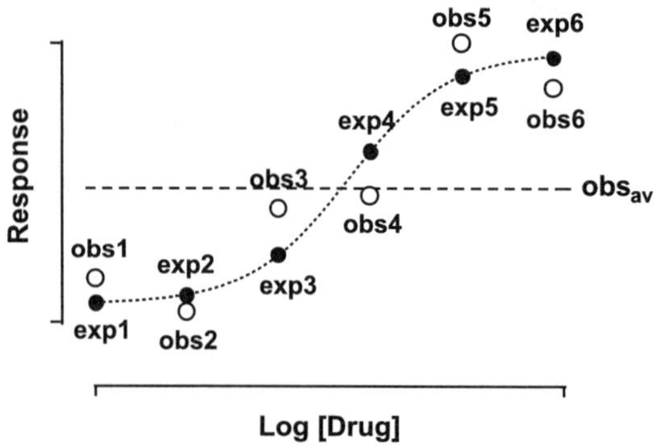

FIGURE 7.10 Relationship between a set of experimental observations (open circles; obs1 – obs6) and their corresponding least squares estimates (solid circles; exp1 – exp6). The horizontal dashed line represents the average of all the experimental observations (obs_{av}).

must also be made for the "degrees of freedom" of the curve fit. For regression procedures, the degrees of freedom equal the total number of datapoints minus the number of model parameters that are estimated. In general, the more parameters that are added to a model, the greater the likelihood of observing a very close fit of the regression curve to the data, and thus a smaller SSE. However, this comes at the cost of degrees of freedom. The "mean square error" (MSE) is defined as the SSE divided by the degrees of freedom (df):

$$\mathrm{MSE} = \frac{\mathrm{SSE}}{\mathrm{df}} \tag{18}$$

Finally, the square root of MSE is equal to the root mean square, RMS:

$$\mathrm{RMS} = \sqrt{\frac{\mathrm{SSE}}{\mathrm{df}}} \tag{19}$$

The RMS (sometimes referred to as $S_{y.x}$) is a measure of the standard deviation of the residuals. It should be noted, however, that although RMS is referred to as the "standard deviation" or "standard error" of the model, this should not be confused with the standard deviation or error associated with the individual parameter estimates. The degree of uncertainty associated with any model parameter is derived by other methods (see below).

7.6.2.3 R^2 (Coefficient of Determination)

Perhaps more common than the RMS, the R^2 value is often used as a measure of goodness of fit. Like the r^2 value from linear regression or correlation analyses, the value of R^2 can range from 0 to 1; the closer to 1 this value is, the closer the model fits the dataset. To understand the derivation of R^2, it is important to first appreciate the other "flavors" of sums-of-squares that crop up in the mathematics of regression procedures in addition to the well-known SSE.

Using Figure 7.10 again as an example, the sum of the squared differences between each observed response and the average of all responses (obs_{av}) is defined as the *Total Sum of Squares* (SST; sometimes denoted as Syy):

$$SST = (obs1 - obs_{av})^2 + (obs2 - obs_{av})^2 + \ldots + (obs6 - obs_{av})^2 \quad (20)$$

where

$$obs_{av} = (obs1 + obs2 + obs3 + obs4 + obs5 + obs6)/6 \quad (21)$$

The sum of the squared differences between each estimated (expected) response, based on the model, and the average of all observed responses is defined as the *Regression Sum of Squares* (SSR):

$$SSR = (exp1 - obs_{av})^2 + (exp2 - obs_{av})^2 + \ldots + (exp6 - obs_{av})^2 \quad (22)$$

The total sum of squares, SST, is equal to the sum of SSR and SSE, and the goal of regression procedures is to minimize SSE (and, as a consequence, SST).

Using the definitions outlined above, the value of R^2 can be calculated as follows[5,10]:

$$R^2 = \frac{SSR}{SST} = 1 - \frac{SSE}{SST} \quad (23)$$

R^2 is the proportion of the adjusted variance in the dependent variables that is attributed to (or explained by) the estimated regression model. Although useful, the R^2 value is often overinterpreted or overutilized as the main factor in the determination of goodness of fit. In general, the more parameters that are added to the model, the closer R^2 will approach a value of 1. It is simply an index of how close the datapoints come to the regression curve, not necessarily an index of the correctness of the model, so while R^2 may be used as a starting point in the assessment of goodness of fit, it should be used in conjunction with other criteria.

7.6.2.4 Analysis of Residuals

Because the goal of least squares regression procedures is to minimize the sum of the squares of the residuals, it is not surprising that methods are available for analyzing the final residuals in order to assess the conformity of the chosen model to the dataset. The most common analysis of residuals relies on the construction of a scatter diagram of the residuals.[13,20]. Residuals are usually plotted as a function of the values of the independent variable. If the model is adequate in describing the behavior of the data, then the residuals plot should show a random scatter of positive and negative residuals about the regression line. If, however, there is a systematic

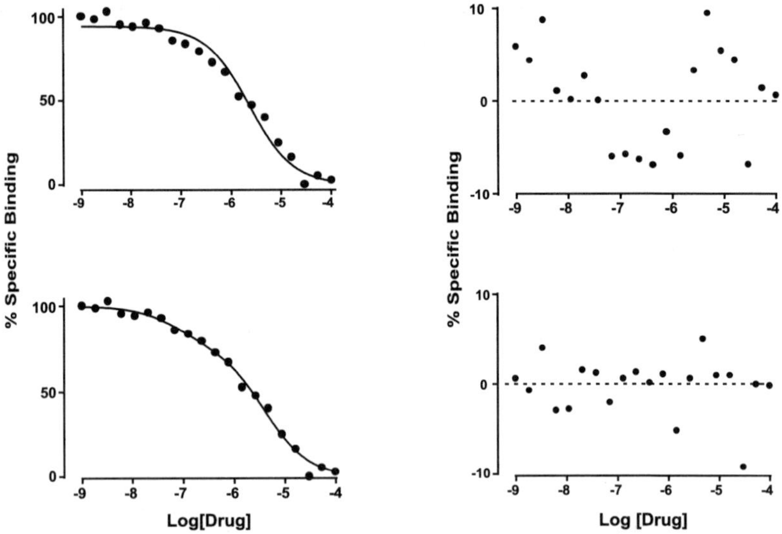

FIGURE 7.11 An example of residuals plots. The top panel represents a curve fit based on a one binding site model to a data set obtained from a radioligand competition binding assay (left) and its corresponding residuals plot (right). Note the clustering of positive and negative residuals. The bottom panel represents a curve fit based on a two binding site model to the same dataset (left) and its corresponding residuals plot (right). Note the random scatter of positive and negative residuals in this case.

deviation of the data from the model, then the residuals plot will show nonrandom clustering of positive and negative residuals. Figure 7.11 illustrates this with an example of a radioligand competition binding experiment. When the data are fitted to a model of binding to a single site, a systematic deviation of the points from the regression curve is manifested as clustering in the residuals plot. In contrast, when the same dataset is fitted to a model of binding to two sites, a random scatter of the residuals about the regression line indicates a better fit of the second model. This type of residual analysis is made more quantitative when used in conjunction with the "runs test" (see below).

There are many other methods of performing detailed analyses of residuals in addition to the common method described above. These methods include cumulative probability distributions of residuals, χ^2 tests, and a variety of tests for serial correlation.[7,10,11,20]

7.6.2.5 The Runs Test

The runs test is used for quantifying trends in residuals, and thus is an additional measure of systematic deviations of the model from the data. A "run" is a consecutive series of residuals of the same sign (positive or negative). The runs test involves a calculation of the expected number of runs, given the total number of residuals and expected variance.[20] The test uses the following two formulae:

$$\text{Expected Runs} = \frac{2N_pN_n}{N_p + N_n} + 1 \tag{24}$$

$$\text{Expected Variance} = \frac{2N_pN_n(2N_pN_n - N_p - N_n)}{(N_p + N_n)^2(N_p + N_n - 1)} \tag{25}$$

where N_p and N_n denote the total number of positive and negative residuals, respectively. The results are used in the determination of a *P* value.[5,13] A low *P* value indicates a systematic deviation of the model from the data. In the example shown in Figure 7.11, the one-site model fit was associated with a *P* value of less than 0.01 (11 runs expected, 4 observed), whereas the two-site model gave a *P* value of 0.4 (10 runs expected, 9 observed).

7.6.3 Optimizing the Fit

With the ubiquitous availability of powerful computers on most desktops, the impressive convergence speed of modern curve fitting programs can often lead to a false sense of security regarding the reliability of the resulting fit. Assuming that the appropriate model has been chosen, there are still a number of matters the biomedical investigator must take into account in order to ensure that the curve fitting procedure will be optimal for their dataset.

7.6.3.1 Data Transformations

Most standard regression techniques assume a Gaussian distribution of experimental uncertainties and also assume that any errors in Y and X are independent. As mentioned earlier, however, these assumptions are not always valid. In particular, the variance in the experimental dataset can be heteroscedastic, that is, it changes in a systematic fashion with the variables. One method for optimizing the curve fitting process to adjust for heteroscedastic errors is to weight the data, as discussed earlier, while another approach is to transform the data to a form where the errors become more homoscedastic prior to the application of the regression technique. Transformations such as the square root or logarithm of the dependent or independent variables do not necessarily cause any problems of their own, provided they reduce rather than increase any heteroscedasticity in the data. In contrast, classical "linearising" transformations, where a new variable is derived from both the original dependent and independent variables, are quite dangerous and it is unfortunate that they are still common practice in some laboratories. Indiscriminate data transforms of the latter kind are troublesome because they have the potential of distorting homoscedastic errors in experimental uncertainties and thus violating the assumptions of any subsequent regression procedure. Transforms are appropriate if they have a normalizing effect on heteroscedastic errors; they are *not* valid otherwise. In addition, some data transforms, embodied in reciprocal plots (e.g., Lineweaver-Burk) or the Scatchard transformation, violate the assumption of independence between X and Y variables

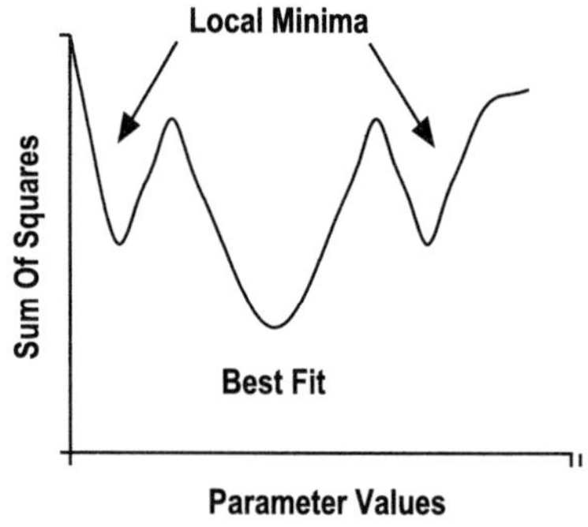

FIGURE 7.12 Multiple minima in parameter space. The best fit is obtained at that set of parameter values yielding the smallest possible sum of squares. Depending on the initial estimates, however, the fitting algorithm may converge on parameter sets which, although yielding a reduced sum of squares, do not correspond to the minimum possible sum of squares. The regression is then said to have converged on a "local minimum."

and are equally inappropriate. In contrast, transformation of model parameters (as described earlier) may often have an optimising effect on the fitting procedure.

7.6.3.2 Initial Estimates

All curve fitting algorithms require the specification of initial estimates of the parameters that are then optimized to yield the best fit. No regression algorithm is perfect, and failure to specify reasonable parameter estimates may result in a failure of the algorithm to converge or, more insidiously, a convergence of the curve fit on a "local minimum." If we recall our earlier discussion of the surface of the merit function that the various algorithms travel down, it is possible to envisage a multi-parameter model that results in a series of troughs such that the algorithm may settle in one as if it has converged on the best fit when, in fact, a deeper trough is available elsewhere on the merit function surface. This is an example of the program converging on a local minimum (Figure 7.12), where the curve fit is not optimal although the user may think that the best fit has been obtained. The best safeguard against this problem is to perform the regression analysis a number of times using different initial estimates. A well-behaved model should converge on essentially the same final estimates each time.

Some commercial programs make the process of finding initial parameter estimates relatively painless by incorporating approximate rules that find initial estimates for the user. Although this is expedient, there is no substitute for the researcher personally addressing the issue of initial parameter estimates. This forces one to focus on the underlying model and the meaning of the model parameters, and it is then not too difficult to come up with a best guess. If further assistance is required, or if there are some parameters that the user does not have a particular "feel" for, then a simplex algorithm or a Monte Carlo-based algorithm (see below) may be utilized to derive estimates that can subsequently be improved upon by the more standard derivative-based algorithms.

7.6.4 Reliability of Parameter Estimates

The determination of the reliability of the estimated parameters derived from a curve fit is as important as the actual estimation of the parametric values themselves. All known methods for the calculation of standard errors and confidence intervals from regression algorithms are based on the mathematics of linear models. Since nonlinear models are more common in biology than linear models, it is perhaps disheartening to have to accept that there are no exact theories for the evaluation of parametric errors in nonlinear regression. However, there are a number of procedures available for approximating these errors such that, in most practical applications, a reasonable measure of parameter error is obtained.

7.6.4.1 Number of Datapoints

The number of experimental datapoints collected and analyzed will play a crucial role in the curve fitting process in one (or both) of two ways:

a. Determination of the appropriateness of the model.
b. Determination of the accuracy of the parameter estimates.

Different measures for goodness-of-fit have already been covered, but some discussion on the influence of datapoint number is also warranted at this point, since it can form an important component of choosing the right model, that adequately accounts for the data. Figure 7.13 illustrates the effect of datapoint number on one of the most common statistical procedures utilized in discriminating between variants of the same model, i.e., the "F-test" (or "extra-sum-of-squares" test). The actual test is described in greater detail in the next section. For now, it is sufficient to point out that the F-test relies heavily on the degrees of freedom associated with the fit to any model, which are in turn dependent on the number of datapoints minus the number of parameters estimated. Although all the points in each of the panels in Figure 7.13 are taken from the same simulated dataset, the "correct" model (a two binding site model) can only be statistically resolved when the datapoints were increased from 6 (panel A) or 10 (panel B), to 20 (panel C).

Assuming that the researcher has *a priori* reasons for deciding that a particular model is most appropriate under their circumstances, the number of datapoints will still be crucial in determining the accuracy of the parameters based on that model. Table 7.1 lists the parameter estimates and corresponding 95% confidence intervals of a two binding site model (i.e., the correct model) applied to the datasets of Panel A and Panel C, respectively, of Figure 7.13. Although the final parameter estimates appear comparable in each instance, the fit based on the small number of datapoints is associated with unacceptably large confidence intervals. There are simply not enough points to accurately define all the parameters of the model. In contrast, increasing the number of datapoints to 20 allowed for reasonable estimates of the error associated with each parameter estimate. The confidence intervals reported in the table were calculated from the asymptotic standard errors derived by the computer program from the fitting algorithm and are most likely *underestimates* of the true error (see below), thus rendering our (already shaken) confidence in the accuracy

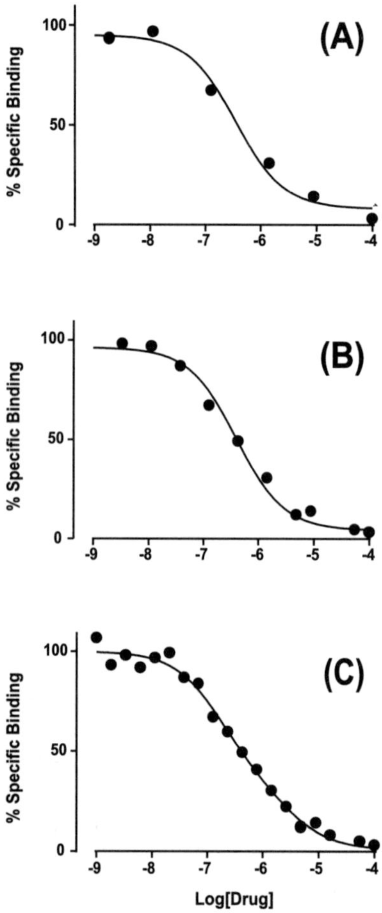

FIGURE 7.13 Influence of data point number on choice of model. The radioligand competition binding curves above were simulated (with random error) according to a model for binding to two-sites. The sampling of points in each of the panels is from exactly the same simulated dataset. The curves in each panel are the least squares fit of the data to either a one- or two-site binding model, as determined by an *F*-test (see Section 7.6.5). Panels A (6 points) and B (10 points) were not statistically significant from a one-site model. Only in panel C (20 points) were the data able to be statistically resolved into the (correct) two-site model fit.

of minimal-data-point parameter estimates virtually nonexistent. There have been some methods presented in the literature for maximizing the reliability of parameter estimates under conditions of minimal datapoint number (e.g., References 21 and 22), but there really is no substitute for a good sampling of experimental datapoints.

7.6.4.2 Parameter Variance Estimates from Repeated Experiments

The most straightforward and conservative approach to building up an error profile of a given parameter is to simply repeat the same experiment many times, obtain

TABLE 7.1
Parameter Estimates and Associated Confidence Intervals from Fitting a Two-Site Model of Radioligand Competition Binding to Different Data Point Numbers Taken from the Same Dataset (Panels A and C; Figure 7.13)

Parameter	Estimate	95% Confidence Interval
	Datapoints = 6	
Maximum Asymptote[a]	–96.7	45.6 to 148.3
Minimum Asymptote[b]	1.3	–84.8 to 87.56
Log IC_{50High}[c]	–6.7	–9.93 to –3.53
Log IC_{50Low}[d]	–5.1	–16.1 to 5.9
Fraction High[e]	0.74	–1.4 to 2.9
	Datapoints = 20	
Maximum Asymptote	99.9	95.4 to 104.4
Minimum Asymptote	0.9	–5.5 to 7.4
Log IC_{50High}	–6.8	–7.3 to –6.5
Log IC_{50Low}	–5.6	–6.4 to –4.6
Fraction High	0.64	0.4 to 0.8

[a] Y-axis value in the absence of competing drug.
[b] Y-axis value in the presence of saturating concentrations of competing drug.
[c] Potency estimate for competition at the high affinity binding site.
[d] Potency estimate for competition at the low affinity binding site.
[e] Fraction of high affinity binding sites.

single parameter estimates from each individual curve fit, and then derive the mean and standard deviation (and error) of the parameters using standard textbook methods. Assuming that each curve fit is performed under optimal conditions, e.g., appropriate number of datapoints, appropriate transformation and weighting, etc., biomedical research is still fraught with small overall sample sizes; it is not uncommon to see n = 3 – 6 given in many publications as the number of times an experiment is repeated. As such, the conservative, albeit straightforward approach to parameter error estimation just described may not have the power to resolve small differences between experimental treatments, as it is based on small sample sizes and, furthermore, does not utilize all the available datapoints. The remaining methods for parameter error estimation utilize all the datapoints in some form or other.

7.6.4.3 Parameter Variance Estimates from Asymptotic Standard Errors

The standard errors reported by practically all commercially available least squares regression programs fall under this category. Asymptotic standard errors are computationally the easiest to determine and, perhaps not surprisingly, the least accurate.

In most instances, these standard errors will underestimate the true error that is likely to be associated with the parameter of interest.

The calculation of the asymptotic standard error and associated confidence intervals involves matrix algebra, but may be summarized as follows:[23]

1. Determine the *Hessian (or "information") matrix.* This is the matrix containing the second derivatives of the parameters with respect to the minimized χ^2 merit function.
2. Evaluate the *variance-covariance matrix* by multiplying the inverse of the Hessian matrix by the variance of the residuals of the curve fit.
3. The diagonal elements of the resulting variance-covariance matrix are the squares of the asymptotic standard errors; the off-diagonal elements of the matrix are the *covariances* of the parameters, and are a measure of the extent to which the parameters in the model are correlated with one another.

The computer program then reports the resulting standard errors. For these errors to actually be a good measure of the accuracy of the parameter estimates, the following assumptions must hold:[23]

a. The fitting equation is linear.
b. The number of datapoints is very large.
c. The covariance terms in the variance-covariance matrix are negligible.

For nonlinear models, the first assumption is invalid, however, the impact of failure to conform to this assumption may be lessened for models that are well behaved, e.g., contain conditionally linear parameters or can be approximated by linear functions. The second assumption can also be reasonable provided the experimenter is able to ensure an adequate sampling of datapoints. Unfortunately, the third assumption is almost never realized. As described earlier, most parameters in nonlinear models show some degree of correlation with one another; indeed, high correlations are indicative of parameter redundancies in the model. As such, ignoring the covariances from the variance-covariance matrix in the reporting of parameter errors will underestimate the true error.

Nevertheless, asymptotic standard errors may serve a useful diagnostic role. Since they will invariably be underestimates of the true error, very large standard errors or confidence intervals reported after a curve fit are indicative of a very poor fit of the associated parameter (see Table 7.1). This may occur, for instance, because the parameter is ill defined by the available data.

7.6.4.4 Monte Carlo Methods

The most reliable method for the determination and validation of model parameter confidence intervals is also the most computer-intensive. Monte Carlo simulations involve the generation of multiple (hundreds to thousands) of pseudodatasets, based on a chosen model, and the subsequent analysis of the simulated datasets with the

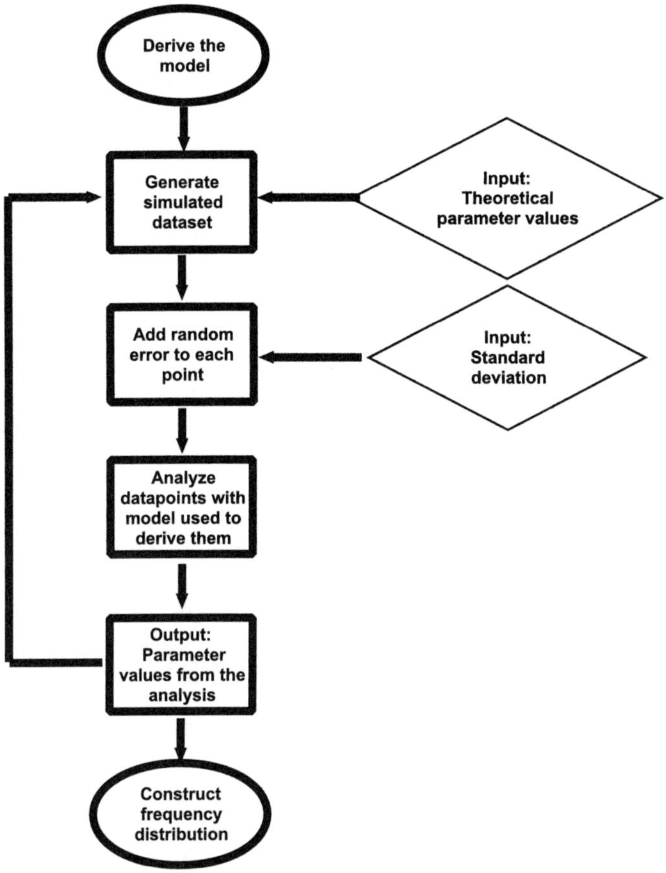

FIGURE 7.14 A general approach to Monte Carlo simulation.

same model used to generate them followed by construction of a frequency histogram showing the distribution of parameter estimates.[17,24] Figure 7.14 shows a flowchart summarizing the general approach to Monte Carlo simulation.

The crucial factor in the implementation of the Monte Carlo approach is the ability to add random "error" to the pseudodataset points that accurately reflects the distribution of experimental uncertainties associated with the determination of "real" datasets. The best determinant of this error is the variance of the fit of the chosen model to real experimental data, provided that the standard assumptions underlying least squares regression analyses are valid. In addition to the appropriate choice of variance for the simulations, other key features in this approach are the choice and the number of independent variables, which again should match those determined in a typical experiment.

The beauty of the Monte Carlo approach is that the level of accuracy with regard to the confidence interval profiles is very much in the hands of the researcher; the greater the number of simulated datasets, the greater the resolution of the confidence intervals. However, this comes at the expense of computer time; a Monte Carlo simulation of 1000 datasets may take 1000 times longer than a least squares fit of the actual experimental dataset used to pattern the simulations. Coupled with the fact that many commercially available curve fitting packages do not contain Monte Carlo-compatible programming features, the time factor involved in generating parameter confidence intervals from the Monte Carlo approach dissuades many researchers from routinely using this method. Nonetheless, great insight can be gained from Monte Carlo approaches. For instance, in addition to providing the greatest degree of accuracy in parameter error estimation, Monte Carlo methods can also guide the experimenter toward the most appropriate model reparameterizations in order to optimize the actual curve fitting procedure.[17]

One potential problem with the standard Monte Carlo approach is that it is necessary to define the population distributions for the errors applied to the datapoints. A normal distribution is most commonly used, but it is not always clear that it is appropriate. The bootstrap, described below, explicitly overcomes that problem.

7.6.4.5 The Bootstrap

"Bootstrapping" is an oddly-named process that allows an approximate reconstruction of the parameters of the population from which the data have been (at least conceptually) sampled.[25] Bootstrapping differs from standard Monte Carlo methods in that it makes no assumption about the form of the population, and instead assumes that the best estimate of the properties of the population is the experimentally determined dataset. The population is reconstructed by repeated resampling of the datapoints to give a large number (hundreds or even thousands) of new pseudodatasets. The resampling is done "with replacement," which is to say that any particular real datapoint can appear in each pseudodataset more than one time. The result is a population of pseudodatasets that represents a pseudo-population that has approximately the same properties as the original population.

Bootstrapping can be used in several ways relevant to model fitting. First, it can provide a pseudopopulation of any parameter calculable from each pseudodataset. Thus it can be used to give confidence intervals for fitted parameters obtained from methods that do not directly provide estimates of parameter variance, such as the simplex method. Similarly, it has been used to estimate the reliability of the variance estimates obtained from other methods that rely on the covariance matrix.[19]

Bootstrapping is not without potential problems. One arises from the fact that the real dataset is unlikely to include any samples from the extreme tails of the overall population of possible datapoints. This means that bootstrapped populations generally have less area under the extreme tails than the real population from which the data were sampled. There are corrections that can be applied,[25] but bootstrapping is not universally accepted by statisticians.

7.6.4.6 Grid Search Methods

Another computer-intensive approach to error determination involves the construction of multidimensional grids based on model parameter values and then "searching" for those parameter value combinations where the variance of the overall fit increases significantly. The confidence intervals are then defined as those regions of the grid (which resemble a multidimensional ellipsoid) that surround the minimum over which the variance does not change significantly.[8,23]

7.6.4.7 Evaluation of Joint Confidence Intervals

As discussed earlier, the parameters in most models tend to show some correlation with one another. The evaluation of joint confidence intervals is a procedure that is designed to include the covariance of the parameters in the determination of parameter error estimates.[8,13] The equations underlying this approach, however, assume that the fitting equation is linear in order to derive a symmetrical elliptically-shaped confidence interval profile of parameters. Unfortunately, this method yields asymmetric confidence regions for those nonlinear models that cannot approximate to a linear model, and is thus not as reliable as Monte Carlo or Grid search methods.

7.2.6.5 Hypothesis Testing

Often, the desire to ascertain the standard error or confidence interval associated with model parameters is a prelude to the statistical testing of the parameters according to a particular hypothesis. Therefore, some objective statistical test is required in order to allow for comparisons between parameters or comparisons between models.

7.6.5.1 Assessing Changes in a Model Fit Between Experimental Treatments

There are three broad approaches to performing statistical comparisons between the same model parameters before and after an experimental treatment. The first relies on the use of standard parametric tests, such as the Student's *t*-test. The second approach relies on more computer-intensive, but preferable comparisons between parameters based on *permutation tests* (see Ludbrook, this volume). The third approach differs from the other two in that it uses all the experimental data generated before and after a particular treatment in a comparison of global changes in goodness of fit. The last procedure may be summarized as follows.[5,26]

1. Analyze each dataset separately.
2. Sum the SSE resulting from each fit to give a new "total" sum-of-squares value (SS_A). Similarly, sum the two degrees of freedom values from each fit to give a "total" degrees of freedom (df_A).
3. Pool the two sets of data into one large set.
4. Analyze this new "global" dataset to obtain a new sum-of-squares value (SS_B) and degrees of freedom (df_B).
5. Calculate the following F ratio:

$$F = \frac{(SS_B - SS_A)/(df_B - df_A)}{SS_A / df_A} \tag{26}$$

The F value is used to obtain a P value, with the numerator having ($df_B - df_A$) degrees of freedom and the denominator having df_A degrees of freedom. A small P value (i.e., large F value) indicates that the individual fits are better than the global, pooled fit i.e., the experimental treatment resulted in a significant difference in the model parameters between the two datasets.

7.6.5.2 Choosing Between Models

The F ratio can also be used to compare the fit of a single dataset to two different versions of the same model:

$$F = \frac{(\mathrm{SS1} - \mathrm{SS2}) / (\mathrm{df1} - \mathrm{df2})}{\mathrm{SS2} / \mathrm{df2}} \tag{27}$$

In this instance, SS1 and df1 are defined as the SSE and degrees of freedom, respectively, of the model with fewer parameters, whereas SS2 and df2 are defined as the SSE and degrees of freedom, respectively, of the model with the greater number of parameters. The addition of more parameters to a model will result in an improvement of the goodness of fit and a reduction in SSE, but at the cost of degrees of freedom. The F test [Equation (27)] attempts to quantify whether the loss of degrees of freedom on going from a simpler to a more complicated model is worth the gain in goodness of fit. A low P value is indicative of the more complicated model being the statistically better model. It should be noted, however, that the F test can only be applied to two different versions of the same model e.g., a one binding-site versus a two binding-site curve fit. In addition, the F test is particularly harsh since it relies so heavily on degrees of freedom and, hence, datapoints and number of parameters. As a consequence, the test may be too conservative and reject the more complicated model for the simpler one, even when this is not the case. Thus, results from the test should be regarded with caution if the number of datapoints is limited and other measures of goodness of fit appear to indicate that the simpler model is not a reasonable fit to the data. When in doubt, repeat the experiment with greater numbers of datapoints.

7.7 FITTING VERSUS SMOOTHING

Throughout this chapter, the process of fitting empirical or mechanistic models to experimental data has generally been encompassed within the umbrella term "curve fitting." However, some distinctions can be made. *Simulation* refers to the process whereby the properties of the model are examined in order to determine the theoretical consequences of imposing specified conditions on the parameters and variables. The term *fitting* refers to the process whereby the model parameters are altered to discover which set of parameter values best approximate a set of experimental observations derived from the actual system of interest. A special case of the fitting process is the procedure known as *smoothing*, whereby a model is chosen to generate a fit that simply passes near or through all the experimental datapoints in order to act as a guide for the eye.

TABLE 7.2
Selected List of Commercially-Available Curve Fitting Programs and Their Associated Least Squares Algorithms. Distributors are listed in parentheses.

Program	Algorithm
Enzfitter (Biosoft)	Levenberg-Marquardt; Simplex
Excel (Microsoft)	Simplex
Fig. P (Biosoft)	Levenberg-Marquardt
Kaleidagraph (Synergy)	Levenberg-Marquardt
KELL (Biosoft)	Levenberg-Marquardt
Origin (Microcal)	Levenberg-Marquardt
Prism (GraphPad)	Levenberg-Marquardt
ProFit (QuantumSoft)	Levenberg-Marquardt; Robust; Monte Carlo (Simplex)
Scientist (Micromath)	Levenberg-Marquardt; Simplex
SigmaPlot (SPSS)	Levenberg-Marquardt

If the purpose of the curve fitting procedure is simply to smooth or to generate a standard curve for extrapolation, then the nature of the underlying model and accompanying regression technique is not crucial. If, however, the purpose of the curve fitting procedure is to obtain insight into the features of the model that describe an aspect of the biological system of interest, then the choice of model is paramount. Although linear models can give curved lines, (e.g., the polynomial equations described earlier), most biological experiments that yield data described by a curve are probably best analyzed using nonlinear regression. This is because it is much more common to find a nonlinear model that can be related in a meaningful and realistic fashion to the system under study than a general linear model.

7.8 CONCLUSION

Computerized curve fitting has become nearly ubiquitous in the analysis of biomedical research. The ease of use and speed of the modern curve fitting programs encourage researchers to use them routinely for obtaining unbiased parameter estimates where in the not very distant past, they might have used eyeballing or linearization processes that would have contained substantial subjective elements and systematic distortions. Nevertheless, indiscriminate use of curve fitting without regard to the underlying features of the model and data is a hazardous approach. We hope that the content of this chapter is useful in illustrating both strengths and some pitfalls of computer-based curve fitting, and some ways to optimize the quality and utility of the parameters so obtained.

7.9 SOFTWARE

Table 7.2 contains a limited sampling of commercially available curve fitting programs. Some of them (e.g., EnzFitter and KELL) are more specialized in their applications than others, but all are commonly applied to curve fitting of biological models to data. Also shown in the table are the associated regression algorithms utilized by each program.

REFERENCES

1. Wells, J.W., Analysis and interpretation of binding at equilibrium, in *Receptor-Ligand Interactions: A Practical Approach,* E.C. Hulme, Ed., Oxford University Press, Oxford, 1992, 289–395.
2. Hill, A.V., The possible effects of the aggregation of the molecules of haemoglobin on its dissociation curves, *J. Physiol.,* 40, iv–vii, 1910.
3. Hill, A.V., The combinations of haemoglobin with oxygen and with carbon monoxide. I, *Biochem. J.,* 7, 471–80, 1913.
4. Motulsky, H.J., *Intuitive Biostatistics,* Oxford University Press, New York, 1995.
5. Motulsky, H.J., *Analyzing Data with GraphPad Prism,* GraphPad Software Inc., San Diego, CA, 1999.
6. Ludbrook, J., Comparing methods of measurements, *Clin. Exp. Pharmacol. Physiol.,* 24(2), 193–203, 1997.
7. Bates, D.M. and Watts, D.G., *Nonlinear Regression Analysis and Its Applications,* Wiley and Sons, New York, 1988.
8. Johnson, M.L. and Faunt, L.M., Parameter estimation by least-squares methods, *Methods Enzymol.,* 210, 1–37, 1992.
9. Press, W.H., Teukolsky, S.A., Vetterling, W.T., and Flannery, B.P., *Numerical Recipes in C. The Art of Scientific Computing,* Cambridge University Press, Cambridge, MA, 1992.
10. Gunst, R.F. and Mason, R.L., *Regression Analysis and Its Applications: A Data Oriented Approach,* Marcel Dekker, New York, 1980.
11. Cornish-Bowden, A., *Analysis of Enzyme Kinetic Data,* Oxford University Press, New York, 1995.
12. Johnson, M.L., Analysis of ligand-binding data with experimental uncertainties in independent variables, *Methods Enzymol.,* 210, 106–17, 1992.
13. Motulsky, H.J. and Ransnas, L.A., Fitting curves to data using nonlinear regression: a practical and nonmathematical review, *FASEB J.,* 1, 365–74, 1987.
14. Jurs, P.C., Curve fitting, in *Computer Software Applications in Chemistry,* Wiley and Sons Inc., New York, 1996, 25–51.
15. Black, J.W. and Leff, P., Operational models of pharmacological agonism, *Proc. Roy. Soc. (Lond.) B.,* 220, 141–62, 1983.
16. Leff, P., Prentice, D.J., Giles, H., Martin, G.R., and Wood, J., Estimation of agonist affinity and efficacy by direct, operational model-fitting, *J. Pharm. Methods,* 23, 225–37, 1990.
17. Christopoulos, A., Assessing the distribution of parameters in models of ligand-receptor interaction: to log or not to log, *Trends Pharmacol. Sci.,* 19, 351–7, 1998.
18. Parker, R.B. and Waud, D.R., Pharmacological estimation of drug-receptor dissociation constants. Statistical evaluation. I. Agonists, *J. Pharmacol. Exp. Ther.,* 177, 1–12, 1971.
19. Lew, M.J. and Angus, J.A., Analysis of competitive agonist-antagonist interactions by nonlinear regression, *Trends Pharmacol. Sci.,* 16(10), 328–37, 1995.

20. Straume, M. and Johnson, M.L., Analysis of residuals: criteria for determining goodness-of-fit, *Methods Enzymol.,* 210, 8–105, 1992.
21. Rovati, G.E., Rodbard, D., and Munson, P.J., DESIGN: computerized optimization of experimental design for estimating K_d and B_{max} in ligand binding experiments. I. Homologous and heterologous binding to one or two classes of sites, *Anal. Biochem.,* 174(2), 636–49, 1988.
22. Rovati, G.E., Rodbard, D., and Munson, P.J., DESIGN: computerized optimization of experimental design for estimating K_d and B_{max} in ligand binding experiments. II. Simultaneous analysis of homologous and heterologous competition curves and analysis of "multiligand" dose-response surfaces, *Anal. Biochem.,* 184(1), 172–83, 1990.
23. Johnson, M.L., Use of least-squares techniques in biochemistry, *Methods Enzymol.,* 240, 1–22, 1994.
24. Straume, M., Veldhuis, J.D., and Johnson, M.L., Model-independent quantification of measurement error: empirical estimation of discrete variance function profiles based on standard curves, *Methods Enzymol.,* 240, 121–50, 1994.
25. Efron, B. and Tibshirani, R.J., *An Introduction to the Bootstrap.*, Vol. 57. Chapman and Hall, New York, 1993.
26. Ratkowsky, D., *Nonlinear Regression Modelling: A Unified and Practical Approach,* Marcel Dekker, New York, 1983.

Index

A

B

C

D

E

J

K

L

M

N

O

P

X

Y

Z